Acoustical Sensing and Imaging

Acoustical Sensing and Imaging

HUA LEE
University of California
Santa Barbara, USA

CRC Press is an imprint of the
Taylor & Francis Group, an **informa** business

CRC Press
Taylor & Francis Group
6000 Broken Sound Parkway NW, Suite 300
Boca Raton, FL 33487-2742

First issued in hardback 2020

First issued in paperback 2023

ISBN-13: 978-1-4987-2573-6 (hbk)
ISBN-13: 978-1-138-58496-9 (pbk)
ISBN-13: 978-0-429-17416-2 (ebk)

DOI: 10.1201/b19578

Publisher's Note
The publisher has gone to great lengths to ensure the quality of this reprint but points out that some imperfections in the original copies may be apparent.

Visit the Taylor & Francis Web site at
http://www.taylorandfrancis.com

and the CRC Press Web site at
http://www.crcpress.com

In memory of Professor Glen Wade

Contents

Preface

The focus of this book is the applications of system analysis and signal processing in acoustical sensing and imaging. The topical coverage of this book is the core component of a graduate-level course on imaging systems. This course has been part of the graduate curriculum in the signal processing area for many years. It was first established in the Electrical and Computer Engineering Department of the University of Illinois at Urbana-Champaign in 1985 and is currently offered in the University of California at Santa Barbara. The 20 sections included in this book represent 20 lectures of the course. The overall objective of this course is to enable students to be familiar with the design concepts, analysis, and development of high-performance sensing and imaging systems with a unified theoretical framework.

For the complex operating modalities and dimensionalities, the design and development of high-performance sensing and imaging systems represents the most direct and significant advances in the field of system analysis and signal processing. In this field, the key components are (1) physical modeling, (2) mathematical analysis, (3) formulation of image reconstruction algorithms, (4) performance evaluation, and (5) system optimization. In the physical modeling of the systems, other than the analysis of the data-acquisition hardware, the focus is the resolution analysis. Instead of the conventional Rayleigh-based formula, to expand the study for systems with discrete distributed arrays as well as various data-acquisition modalities, the resolution analysis is conducted here through the quantitative assessment of the spatial-frequency spectral coverage. In the area of signal processing, the key elements are the design of the probing waveforms, image reconstruction algorithms, error reduction and removal, and image enhancement. With a unified framework, the image reconstruction algorithms are formulated based on the concept of coherent backward propagation, in the form of multi-frequency tomography. For the improvement of system performance, the main topics are the correction of quadrature phase errors prior to image reconstruction and enhancement with coherent wavefield statistics during the superposition of sub-images.

The most effective summary of the design and development of high-performance sensing and imaging systems is the direct applications. In this book, several applications have been included as examples of various operating modalities. These are results from the research projects conducted in the Imaging Systems Laboratory during the past 30 years. These projects fully illustrate the technical and educational significance of this field. I wish to take this opportunity to acknowledge the generous support of these research programs provided by the National Science Foundation, the Defense Advanced Research Projects Agency (DARPA), Army Research

Office (ARO), the University of California Microelectronics Innovation & Computer (UC MICRO) Program, the U.S. Department of Energy, the U.S. Department of Transportation, the W. M. Keck Foundation, the International Foundation of Telemetering, 3M, Sonatech, the Rockwell International, and the D'Errico research funds.

I would also like to thank the members of my research laboratory, Jason Lin, Richard Chiao, Brett Douglas, John Fumo, Davis Kent, Mei-Su Wu, Daniel Doonan, Christopher Utley, and Michael Lee, who executed these acoustical imaging research projects with extraordinary dedication and enthusiasm. Finally, I am especially grateful to my PhD thesis advisor, Professor Glen Wade, for his vision of this important and exciting research field.

Hua Lee
University of California, Santa Barbara
2015

Author

Professor Hua Lee earned his BS from the National Taiwan University, Taipei, Taiwan, in 1974, and MS and PhD from University of California, Santa Barbara (UCSB), California, in 1978 and 1980, respectively. He returned to UCSB in 1990 where he is currently a professor of the Electrical and Computer Engineering Department. Prior to his return to UCSB, he was on the faculty of the University of Illinois at Urbana-Champaign, Illinois. His research interests cover the areas of imaging system optimization, high-performance image formation algorithms, synthetic aperture radar and sonar systems, acoustic microscopy, microwave nondestructive evaluation, terahertz imaging, tomographic ground-penetrating radar imaging, and reconfigurable sensing systems.

Professor Lee served as department vice chair from 1993 to 1995 and department chairman from 1998 to 2002. From 2007 to 2012, he served as technical director of the National Security Institute. In the professional societies, Dr. Lee served as the chairman of the 18th, 24th, and 30th International Symposium on Acoustical Imaging in 1989, 1998, and 2009, respectively. He also served as co-chair of the 13th International Workshop on Maximum Entropy and Bayesian Methods in 1993 and the 9th International Ground-Penetrating Radar Conference in 2002. From 1988 to 1994, he served as editor of the *International Journal of Imaging Systems and Technology*. He also served as an associate editor of *IEEE Transactions on Circuits and Systems for Video Technology* from 1992 to 1995, and associate editor of *IEEE Transactions on Image Processing* from 1994 to 1998.

Professor Lee received the Presidential Young Investigator Award in 1985 for his work in imaging system optimization. He was elected *Professor of the Year* in 1992 by the Mortar Board National Honor Society. Dr. Lee was also given the Nineteenth Pattern Recognition Society Award in 1993. In 1998, he received the UCSB Academic Senate Distinguished Teaching Award. He also received the UCSB College of Engineering Outstanding Faculty Award in 2000, 2006, and 2015. In 2011, he received the Technical Achievement Award of the International Symposium on Acoustical Imaging. Professor Lee is a Fellow of the Acoustical Society of America and IEEE.

Dr. Lee's publications include the books *Imaging Technology* (IEEE Press, 1986), *Modern Acoustical Imaging* (IEEE Press, 1986), *Engineering Analysis: A Vector Space Approach* (John Wiley, 1988), *Acoustical Imaging*, Vol. 18 (Plenum Press, 1991), *Acoustical Imaging*, Vol. 24 (Kluwer Academic/Plenum Publishers, 2001), *Acoustical Imaging*, Vol. 30 (Kluwer Academic/Plenum Publishers, 2011), and *Biomedical Devices and Technology* (Wiley, 2012).

1

Introduction

The purpose of this book is to provide a comprehensive overview of all the technical aspects in acoustical sensing and imaging. The coverage includes system modeling, resolution analysis, signal processing, error reduction, resolution enhancement, and motion estimation, with results from full-scale field experiments and applications. This book can be used by engineers for a concise review of the technical elements of this field. It can also serve as the textbook for a senior-elective or a first-year graduate course in electrical engineering.

Chapter 2 focuses on the topic of bearing-angle estimation with underwater acoustic sensors. It consists of two parts. The first part is based on the design of multi-element transmitters and one single receiver. This version is to minimize the computation complexity and power consumption of the receiver of the mobile unit. A simple high-performance signal processing technique for the estimation of the bearing angle provides excellent stability, accuracy, and computation efficiency. The utilization of the orthogonality of Hilbert transform pairs is then introduced for the optimization of the transmission waveform design. The second version is based on the design of one single transmitter and a receiver with multiple elements. In terms of both hardware design as well as signal processing, these two versions have a high level of symmetry and duality. This version further simplifies the hardware electronics of the transmission operations. There are several potential extensions of this design concept. One is to modify the system into an active system with the addition of transmitter at the center of the receiver array elements. This enables the system to gain range-estimation capability such that we can fully determine the location of the target. Another extension is the feasibility of nonuniform receiver array elements. It allows us to conduct bearing-angle estimation without the requirement of structured array configurations.

Chapter 3 is the core theoretical component of the book. Its key objective is to model the wave propagation and scattering as a multi-dimensional linear and shift-invariant operation. This model allows us to apply well-established concepts to the signal analysis and processing for image reconstruction as the inverse-filtering process. The linear modeling includes the formulation of the multi-dimensional transfer function, transfer function of the plane-to-plane model, and Fresnel and Fraunhofer approximations. This leads to the formulation of the backward-propagation technique as the foundation of image reconstruction. Subsequently, the system analysis is followed by the overview of holographic and tomographic imaging modality. A full-scale resolution analysis is also included to formulate the resolution limit of modern systems with discrete sensor arrays.

The main objective of Chapter 4 is the utilization of theoretical concepts and formulations for practical applications. It starts with the applications in acoustic microscopy with the coherent backward-propagation image formation technique. Subsequently, the pulse-echo imaging modality is introduced for the increase of bandwidth to achieve high resolution in the range direction. Synthetic-aperture sonar imaging is used as a direct example.

For the improved utilization of the bandwidth, linear chirp and step-frequency FMCW (frequency-modulated continuous wave) waveforms are then introduced as the probing signals. This leads to the application in high-resolution ultrasound imaging with step-frequency FMCW waveforms. In addition, in this chapter, the reversed version of the multi-frequency backward-propagation imaging technique is used to illustrate the concept of imaging with reconfigurable arrays.

In the field of imaging technology, many important tasks are subsequent to the image reconstruction. This process allows us to conduct thorough system analysis and performance evaluation for modification and optimization. To illustrate this important aspect, Chapter 5 covers four subjects, grouped into two parts. The first part is devoted to enhancement techniques. The first is for the reduction or removal of phase errors introduced to the data-acquisition process for the use of quadrature receivers. The second technique is the formation of an enhancement operator with wavefield statistics, based on the fundamental concept of backward-propagation image formation, for resolution improvement. The second half of the chapter is allocated to motion estimation techniques. It covers both the parameter-based and image-based techniques in both space and frequency domain. In addition, imaging of targets with micro-periodic motions is discussed.

2

Underwater Signal Parameter Detection and Estimation

This chapter presents a simple high-performance technique for underwater geolocation and navigation. The simplicity of this technique provides excellent stability, accuracy, and computation efficiency. In addition, the chapter also describes the reversed version of the system for the enhancement of system performance, and the extension to active system for unmanned underwater vehicle (UUV) collision avoidance.

The original design goal was to improve the accuracy and stability of the conventional polarity estimation algorithm for UUV homing and docking exercises underwater. The assumption is that the relative left–right position of the UUV to the base stations can be estimated from the polarity of the beacon waveforms. The traditional technique conducts the polarity estimation based on the phase term of the first peak of the received beacon signal. In the underwater environment, this approach has been problematic with low stability and accuracy, because of the difficulty especially in locating the first peak from the interference signal pattern under substantial multipath and background noise. After a series of experiments for improving existing algorithms, the effort was then redirected toward the design and development of new techniques with simple structures and computation efficiency. As a result, the *double-integration method* was developed and the results showed superior stability and accuracy.

In the analysis, it was also noted that the reversed version of the existing system could be even more effective. This involves the reversal of the configuration of the transmitter and receiver. The alternative arrangement further simplifies the hardware and software. Besides, the new configuration enables us to continue the improvement of system performance with added receiver elements without increasing the computation complexity.

One important extension is the conversion of the existing system to the active modality by placing a transmitter at the center of the circular receiver array. Because the transmitted signal is available as the reference waveform, this enables us to integrate the estimation of the range distance into the algorithm. For the superior computation efficiency in dynamic sensing, this system will be an excellent candidate for added applications to collision avoidance.

2.1 Acoustic Sensor Unit

In mobile autonomous sensing, the geolocation capability for navigation and guidance can be achieved by estimating the position of the sensor unit with respect to a collection of underwater base stations. The locations of the underwater base systems are constantly estimated and updated with respect to the interface stations over the ocean surface, and the interface stations are supported by the GPS systems with the direct microwave link to surface communication infrastructure. Thus, one of the most critical technical elements is the mobile sensor's capability of dynamic estimation and updating of its relative position with respect to the underwater base stations.

In three-dimensional underwater geolocation tasks, the objective parameters include mainly the *range distance* and a multi-dimensional *bearing-angle vector*. This chapter focuses on the estimation of the relative bearing angle of the transmitter position, before the extension to include the estimation of range distance.

Figure 2.1 shows the laboratory prototype of the three-component transmitter of the base stations. Each of the three components consists of four small square transmitting elements. Figure 2.2 shows the single-element receiver at the mobile platform.

The final prototype design contains only one set of 4-element transmitters, instead of three sets. The preliminary prototype for the laboratory experiments uses the three-transmitter version, for simplicity of the hardware electronics.

With the three transmission components, the transmitter unit sends out a sequence of three signals, $\{T_1(t), T_2(t), T_3(t)\}$. The first signal $T_1(t)$, from the top transducer element, is an in-phase reference signal, of which all four elements send out the same signal. The second signal $T_2(t)$, from the middle

FIGURE 2.1
Laboratory prototype of the three-component transmitter of the base stations.

FIGURE 2.2
Front and back of the single-element receiver.

transducer, is a beacon signal, transmitting a pair of signals with a left–right polarity of 180° phase offset, for the estimation of the bearing angle in the horizontal direction. Similarly, the third signal $T_3(t)$, from the bottom transducers, is with a top–bottom polarity of a 180° phase offset for the estimation of the bearing angle in the vertical direction.

Figure 2.3 shows the polarity distributions of the three sets of transmitters. The left configuration is of the calibration signal from the top transmitter with all four elements in phase. The middle is with left–right polarity of 180° phase offset, and the right is with a top–bottom polarity of 180° phase offset.

As shown in Figure 2.4, the relative time offset of the twin transmitted waveforms is a function of the bearing angle θ and the separation distance D.

Because of the complexity and sensitivity of the underwater acoustic propagation and serious multipath interference, the accuracy of geolocation has been an extremely difficult problem. For the estimation of polarity, the conventional approach is to search for the first peak of the matched-filtered received signal and identify its phase term. The approach encountered several technical difficulties. One is the determination of the location of the first peak, followed by the estimation of it phase term. Secondly, the required computation impacted the complexity of electronic components, latency, as well as the power consumption. Therefore, this procedure has long been regarded as the bottleneck of high-precision underwater geolocation and navigation.

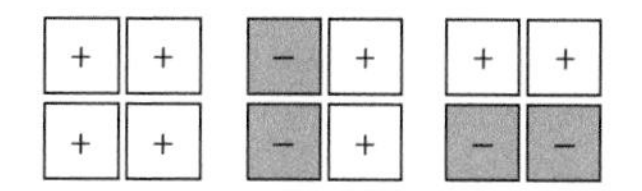

FIGURE 2.3
Polarity distributions of the three sets of transmitters.

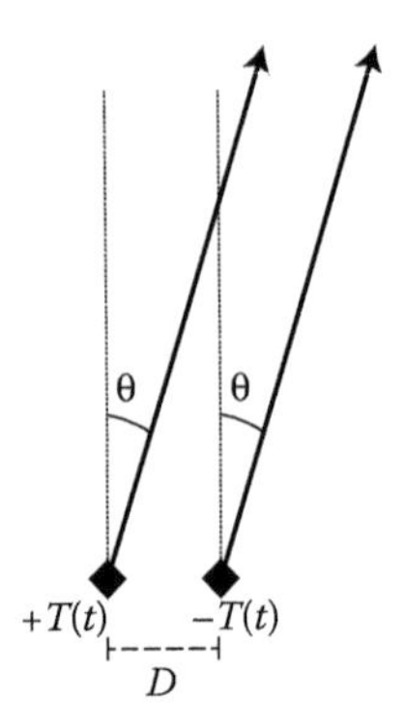

FIGURE 2.4
Relative time offset of the twin transmitted waveforms.

2.2 Direct-Integration Method

To start the analysis, the basic structure of the received beacon signal is examined.

From a particular receiver position, the twin beacon signal, with the polarity of 180° phase offset, from the transmitter array can be written in the form of

$$T_2(t) = \left[h\left(t+\frac{\Delta}{2}\right) - h\left(t-\frac{\Delta}{2}\right)\right] = \left[\delta\left(t+\frac{\Delta}{2}\right) - \delta\left(t-\frac{\Delta}{2}\right)\right] * h(t)$$

where $h(t)$ is the designated transmission waveform for the horizontal direction. For the laboratory prototype, the signal $h(t)$ has a carrier frequency of 73.9 kHz with a 15.8 kHz bandwidth. The positive-polarity term $h(t + \Delta/2)$ is transmitted from one element and the negative-polarity term $-h(t - \Delta/2)$ is from the other transmitter element.

The time-delay term Δ is the relative propagation lag due to the separation distance D between the transmitter elements:

$$\Delta = \frac{D}{v}\sin(\theta)$$

where θ is the unknown bearing angle and v is the propagation speed. The width of the separation D is 1 cm.

It should be pointed out that the relative time-shift term Δ can be either positive or negative, depending on the value of the bearing angle. If the receiver is at the right-hand side of the transmitter, the angle θ is positive and Δ becomes a positive value accordingly. Then the positive-polarity term $\delta(t + \Delta/2)$ arrives at the receiver first, followed by the negative-polarity term

$-\delta(t - \Delta/2)$, Δ seconds later. On the other hand, when the receiver is at the left-hand side of the transmitter, the angle θ becomes negative, and so is Δ. Then the negative-polarity term, $-\delta(t - \Delta/2)$ arrives at the receiver first, followed by the positive-polarity term $\delta(t + \Delta/2)$.

The main objective of the algorithm is the accurate estimation of the term Δ, which gives the estimate of the polarity as well as the bearing angle. At the receiver, the detected signal at the receiver is in the form of

$$s(t) = T_2(t-d) = c\left[\delta\left(t-d+\frac{\Delta}{2}\right) - \delta\left(t-d-\frac{\Delta}{2}\right)\right] * h(t)$$

where d is the time delay due to the propagation from the center of the transmitter-array unit to the receiver, and c is the attenuation factor due to propagation loss.

At the receiving end, after the demodulation for the removal of the carrier frequency, matched filtering is applied to the received signal. Mathematically, the matched filtering process can be formulated in the form of a convolution with $h^*(-t)$.

$$r_0(t) = s(t) * h^*(-t) = c\left[\delta\left(t-d+\frac{\Delta}{2}\right) - \delta\left(t-d-\frac{\Delta}{2}\right)\right] * [h(t) * h^*(-t)]$$

$$= c\left[\delta\left(t-d+\frac{\Delta}{2}\right) - \delta\left(t-d-\frac{\Delta}{2}\right)\right] * R_0(t)$$

$$= c\left[R_0\left(t-d+\frac{\Delta}{2}\right) - R_0\left(t-d-\frac{\Delta}{2}\right)\right]$$

where $R_0(t)$ is the autocorrelation of $h(t)$. Figure 2.5 shows the autocorrelation function $R_0(t)$ of the prototype used in the field experiment.

Because $R_0(t)$ is zero at $t = \pm\infty$, the initial and final values of the matched-filtered signal $r_0(t)$ are zero,

$$r_0(-\infty) = r_0(\infty) = 0$$

Figure 2.6 shows a typical received signal from the laboratory experiment, after matched filtering. The degree of fluctuation of the signal also explains the difficulty in accurate estimating the bearing angle with the conventional approach.

Now, we introduce a new function $r_1(t)$ by directly integrating $r_0(t)$,

$$r_1(t) = \int_{-\infty}^{t} r_0(\tau)\,d\tau = c\left[u\left(t-d+\frac{\Delta}{2}\right) - u\left(t-d-\frac{\Delta}{2}\right)\right] * R_0(t) \pm cp_\Delta(t-d) * R_0(t)$$

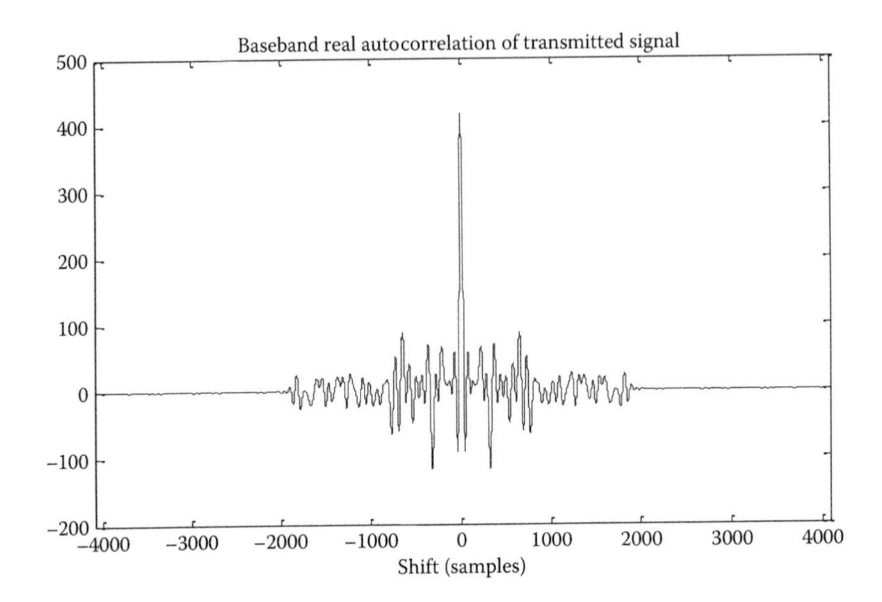

FIGURE 2.5
Autocorrelation function $R_0(t)$ used in the field experiment.

where $p_\Delta(t)$ is a pulse of unit amplitude with pulse duration $|\Delta|$. Note that the duration of the pulse is independent of the time-delay term d. The $\pm$ polarity of the term depends solely on the value of Δ. And the polarity of the term follows exactly the sign of the time delay Δ.

This equation also shows $r_1(t)$ is the result of a convolution of the autocorrelation function $R_0(t)$ with a finite-length pulse with a $\pm$ polarity and length $|\Delta|$. If we perform the second integration, it results in

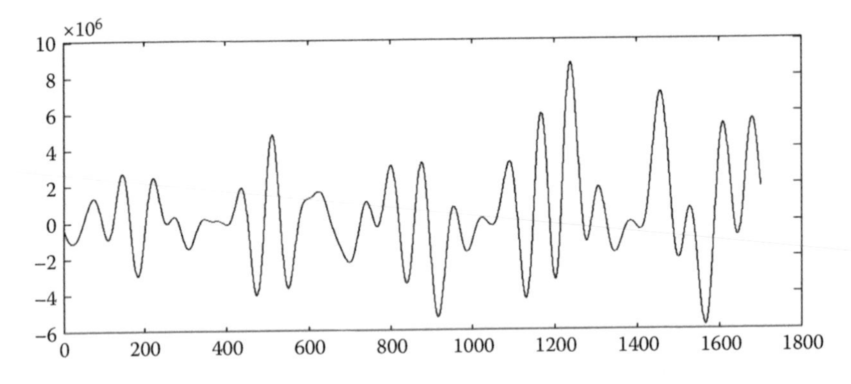

FIGURE 2.6
A received signal after matched filtering.

$$r_2(t) = \int_{-\infty}^{t} r_1(\tau)d\tau = \pm c p_\Delta(t-d) * \int_{-\infty}^{t} R_0(\tau)\, d\tau = \pm c p_\Delta(t-d) * R_1(t)$$

where $R_1(t)$ is the result of integration of $R_0(t)$,

$$R_1(t) = \int_{-\infty}^{t} R_0(\tau)\, d\tau$$

Thus, the final value of $R_1(t)$ is the DC term of $R_0(t)$. Since $R_0(t)$ is the autocorrelation of $h(t)$, the DC term of $R_0(t)$ is $|H(0)|^2$, where $H(j\omega)$ is the Fourier spectrum of $h(t)$. Hence,

$$R_1(\infty) = |H(0)|^2$$

Subsequently, we find the final value of $r_2(t)$ as

$$\text{Final value} = r_2(\infty) = c\Delta \cdot R_1(\infty) = c\Delta \cdot |H(0)|^2 = c\frac{D\sin(\theta)}{v}|H(0)|^2$$

$$= \frac{cD}{v}|H(0)|^2 \sin(\theta)$$

It should be noted that the term, $|H(0)|^2$, is the DC term of the power spectrum of the transmitted signal $h(t)$. Thus, the separation of the transmitter elements D, propagation velocity v, propagation attenuation c, and the term $|H(0)|^2$ are all known parameters. Figure 2.7 summarizes the computational procedure of the estimation process with one set of beacon waveforms.

As a result, the final value of $r_2(t)$ is linearly related to $\sin(\theta)$. This means that the final value of $r_2(t)$ can uniquely characterize the polarity of the signal as well as the bearing angle of the receiver with respect to the transmitter. Figure 2.8 is the result of the double integration with the data from laboratory experiments, showing steady convergence to a collection of values in proportion to $\sin(\theta)$ of the bearing angles.

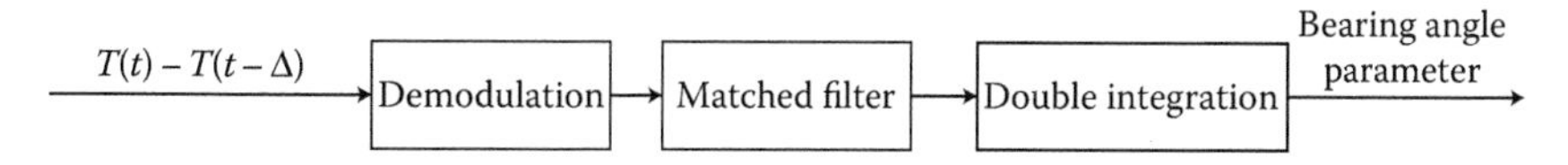

FIGURE 2.7
Computational procedure of the estimation process.

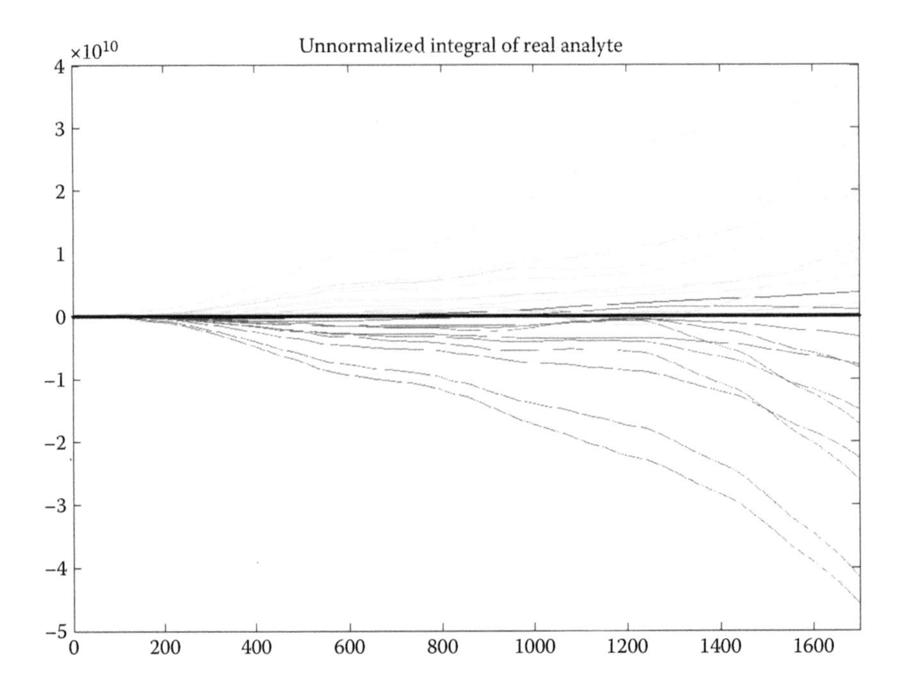

FIGURE 2.8
Results of the double-integration algorithm.

2.3 Normalization and Optimization

The parameters in the constant term, D, v, and $|H(0)|^2$, are fully defined by the system design. The propagation-loss term c is the only uncertain element. The propagation loss is a function of the range distance and beam patterns of the trasmitters and receiver. When the position of the mobile sensor unit changes, the range distance and beam pattern change and the value of the c term also varies as a result, which introduces ambiguity to the estimation of the bearing angle. Thus, to achieve high-precision estimation of the bearing angle, the constant c needs to be isolated from the estimation process.

The calibration signal, the first pulse from the transmitter array, is in the form of

$$T_1(t) = \left[h\left(t + \frac{\Delta}{2}\right) + h\left(t - \frac{\Delta}{2}\right)\right] = \left[\delta\left(t + \frac{\Delta}{2}\right) + \delta\left(t - \frac{\Delta}{2}\right)\right] * h(t)$$

The corresponding detected signal at the receiver becomes

$$s(t) = T_1(t-d) = c\left[\delta\left(t-d+\frac{\Delta}{2}\right)+\delta\left(t-d-\frac{\Delta}{2}\right)\right] * h(t)$$

After matched filter, the signal is

$$r_0(t) = s(t) * h^*(-t) = c\left[\delta\left(t-d+\frac{\Delta}{2}\right)+\delta\left(t-d-\frac{\Delta}{2}\right)\right] * R_0(t)$$

$$= c\left[R_0\left(t-d+\frac{\Delta}{2}\right)+R_0\left(t-d-\frac{\Delta}{2}\right)\right]$$

After one integration, it becomes

$$r_1(t) = \int_{-\infty}^{t} r_0(\tau)\, d\tau = c\int_{-\infty}^{t} R_0\left(\tau-d+\frac{\Delta}{2}\right)+R_0\left(\tau-d-\frac{\Delta}{2}\right) d\tau$$

$$= c\left[R_1\left(t-d+\frac{\Delta}{2}\right)+R_1\left(t-d-\frac{\Delta}{2}\right)\right]$$

Then, the final value of $r_1(t)$ is

$$\text{Final value} = r_1(\infty) = 2cR_1(\infty) = 2c|H(0)|^2$$

Now, it can be seen that the ratio of the final values of the double integration of the beacon signal and the first integration of the reference waveform is

$$\rho = \frac{D}{2v}\sin(\theta)$$

The normalization effectively removes the ambiguity factor due to propagation loss and as a result isolates the bearing-angle estimation from sources of errors. Since the separation distance D and propagation speed v are known, the bearing angle can be estimated with high accuracy. Figure 2.9 summarizes the normalization process with the calibration signal.

This algorithm was applied to the same data set and the result is documented in Figure 2.10, which shows superior performance and accuracy. The top figure shows the result of the double-integration algorithm without the normalization procedure, and the bottom figure is the result after normalization.

Ideally, based on the design of the experiments, the estimates of the bearing angles would be a linear function, from –30° to +30°. The deviation of the result from the line represents the estimation errors of the algorithm. From

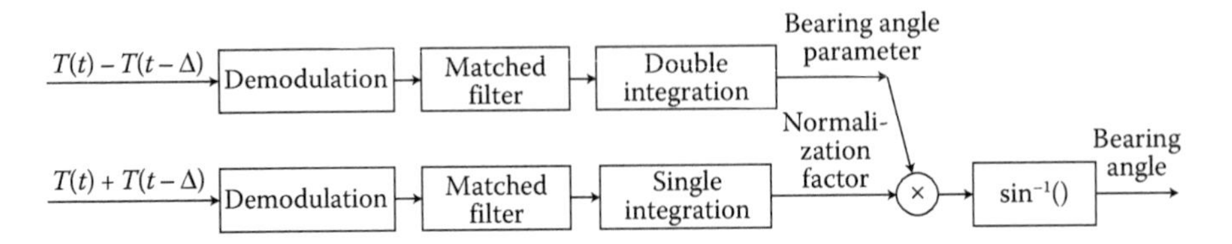

FIGURE 2.9
Signal normalization procedure.

the figure, the distributions of the estimated bearing angles seem similar and in proportion. Note that the scale of the top figure is 10^4 times of that of the bottom figure. Thus, from the scales of the two plots, the rescaling by the normalization process significantly reduces the estimation errors. Figure 2.11 shows the estimation of the bearing angles after the nonlinear conversion.

$$\theta = \sin^{-1}\left(\frac{2v\rho}{D}\right)$$

The accuracy at the edges of the curve, corresponding to the angles at ±30°, is mainly due to the beam patterns of the transceivers and the noise is amplified by the nonlinear *arcsine* operation in the conversion.

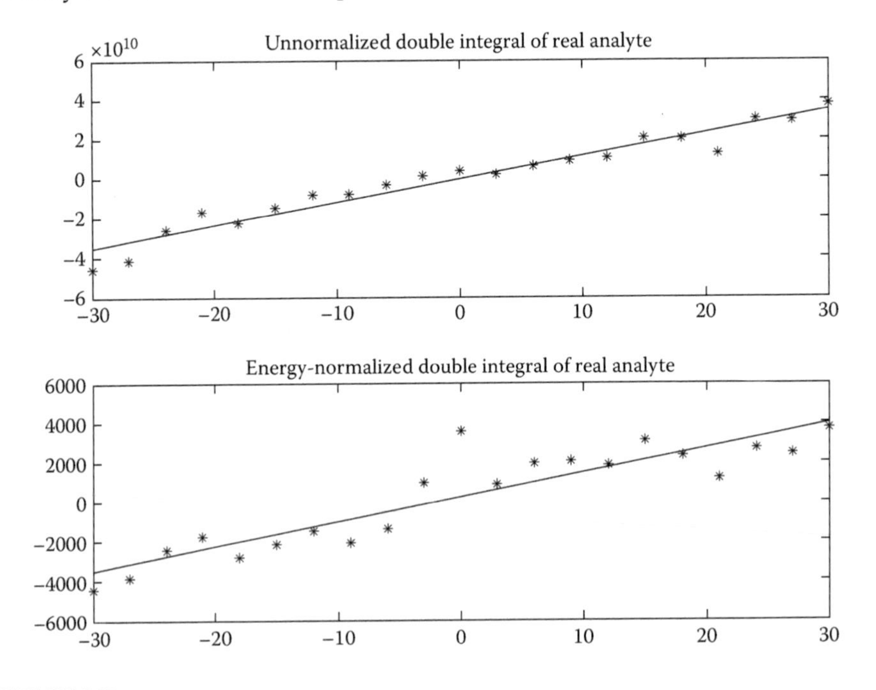

FIGURE 2.10
Estimation of the bearing angles before the normalization (top) and after the normalization (bottom).

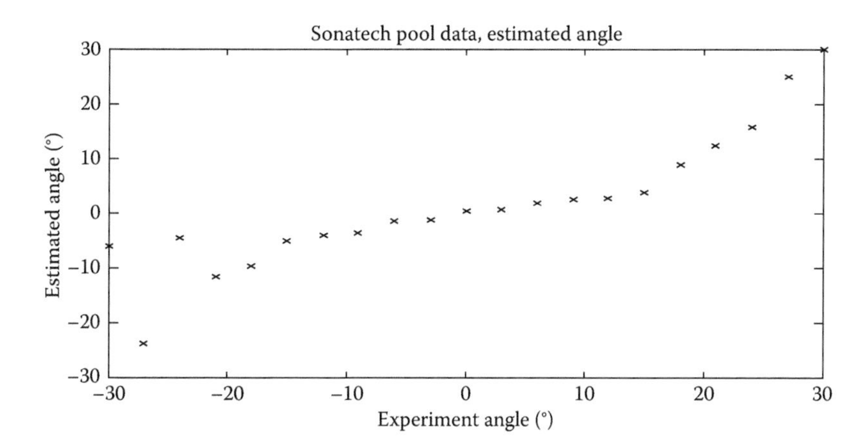

FIGURE 2.11
Estimation of the bearing angles after the nonlinear conversion.

2.3.1 System Optimization

The signal transmission procedure of the preliminary prototype is to transmit a sequence of three separate waveforms $T_1(t)$, $T_2(t)$, and $T_3(t)$, for calibration, and the estimation of horizontal and vertical bearing angles, respectively. To enhance the efficiency and effectiveness of the system, it is ideal to combine the two bearing-angle operations into one. To maximize the efficiency and minimize the coupling, the transmitted signals, $h_x(t)$ and $h_y(t)$, can be organized in the form of a Hilbert transform pair.

$$h_y(t) = H\{h_x(t)\}$$

where $h_x(t)$ and $h_y(t)$ are the beacon signals for the estimation of horizontal and vertical bearing angles, respectively. Because they are a Hilbert transform pair, they are mutually orthogonal.

$$\langle h_x(t), h_y(t) \rangle = 0$$

The signals, $h_x(t)$ and $h_y(t)$, operate in the identical frequency band and have the same power spectrum. The orthogonality removes the leakage during the matched filtering process and improves the accuracy of the overall estimation process.

The Hilbert transform can be characterized as a linear and time-invariant filter with a 90° phase shift. Thus, the phase shift of the four elements of the circular transmitter array can be written as

$$\left[\exp(j0), \exp\left(\frac{j\pi}{2}\right), \exp(j\pi), \exp\left(\frac{j3\pi}{2}\right)\right] = [1, j, -1, -j]$$

This is equivalent to multiplying the signals from the four transmitter elements with the complex weighting of [1, j, –1, –j]. The elements [1, –1] are the weighting coefficients for the two elements in the horizontal direction and [j, –j] are for the vertical direction. In practice, it can be implemented very effectively by shifting the phase term of the carrier waveform,

$$h_x(t) = h(t)\cos(\omega_o t)$$

and

$$h_y(t) = h(t)\sin(\omega_o t)$$

where ω_o is the carrier frequency. The 90° phase shift will be embedded into the signal automatically through the demodulation process. Thus, this optimization procedure allows us to utilize a simple phase shift of the carrier to significantly enhance the performance of the system. The same beacon waveform $h(t)$ is applied, so that the signal generation and matched filtering processes remain the same, without increasing the complexity of the hardware. The signals from the 4-element array, $\{h_x(t), h_y(t), -h_x(t), -h_y(t)\}$, are transmitted simultaneously, so that the data acquisition and processing time can be allocated more effectively. This is to modify the 4-element transmitter for the transmission of four phase-offset waveforms in the form of

$$h_0(t) = h_x(t) = +h(t)\cos(\omega_o t)$$
$$h_1(t) = h_y(t) = +h(t)\sin(\omega_o t)$$
$$h_2(t) = -h_x(t) = -h(t)\cos(\omega_o t)$$
$$h_3(t) = -h_y(t) = -h(t)\sin(\omega_o t)$$

Figure 2.12 shows the geometric arrangement of the 4-element transmitter unit based on the concept of Hilbert transform pairs. It utilizes the

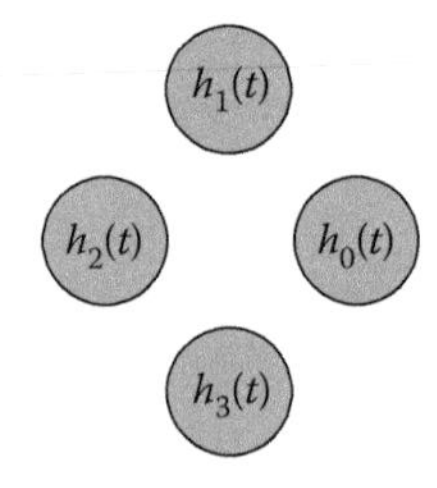

FIGURE 2.12
Geometry of the 4-element transmitter unit.

orthogonality for the estimation of the bearing angles in the horizontal and vertical direction in the combined format.

To improve the accuracy, one option is to increase the number of the array elements. Suppose there are N elements along the circular transmitter array. The phase shifts associated with the array elements are

$$\left\{\exp\left(\frac{j2\pi n}{N}\right)\right\} = \left[\exp(j0), \exp\left(\frac{j2\pi}{N}\right), \exp\left(\frac{j4\pi}{N}\right), \ldots, \exp\left(\frac{j2\pi(N-1)}{N}\right)\right]$$

for $n = 0, 1, 2, \ldots, N-1$. The practical implementation of this concept can be achieved by shifting the carrier waveform with the corresponding phase term,

$$h_n(t) = h(t)m_n(t)$$

where the carrier signals at various transmitting elements are

$$m_n(t) = \cos\left(\omega_o t + \frac{2\pi n}{N}\right) = \cos\left(\frac{2\pi n}{N}\right)\cos(\omega_o t) - \sin\left(\frac{2\pi n}{N}\right)\sin(\omega_o t)$$

This modification significantly improves the overall performance of the system with simple adjustments to the phase of the carrier waveforms. The beacon signal and operating frequency remain exactly the same. Figure 2.13 shows the geometric arrangement of the 8-element transmitter unit, based on the same concept. The received signal is the superposition of the eight waveforms. The computational complexity remains exactly the same, with (a) demodulation, (b) matched filtering, and (c) signal integration. Thus, the improvement of the system design does not introduce any increase of computation complexity, in the form of either hardware or software.

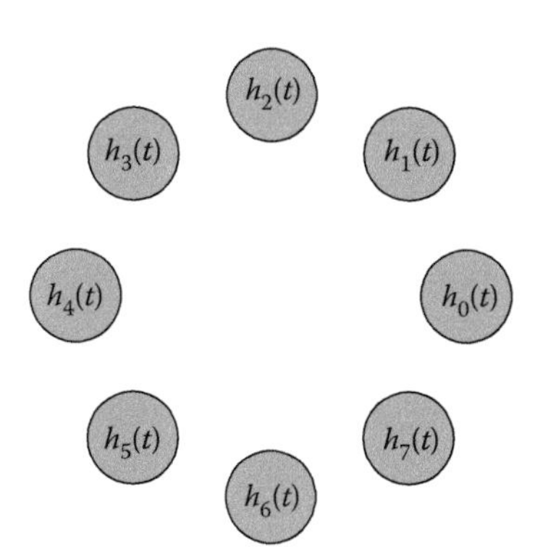

FIGURE 2.13
Geometry of the 8-element transmitter unit.

2.4 Reversed Configuration

It can be seen that the acoustical geolocation tasks can function even more effectively with the reversed version of the system. This is to use a single-element transmitter and a multiple-element receiver array. The reversed version simplifies the transmission operation so that it is required to transmit only one signal instead of a sequence of three. It also makes the transmitter electronics more energy efficient.

Suppose the receiver is a 4-element circular array. The diameter of the array is kept at D similar to the prior setup. The received, demodulated, and matched-filtered signals from the four receiver elements are denoted $\{r_{+x}(t), r_{+y}(t), r_{-x}(t), r_{-y}(t)\}$, respectively.

We can first obtain the normalization factor as the final value of the integral of the four signals. Subsequently, the four signals are normalized to remove the variation due to beam patterns, propagation loss, and transducer characteristics. Then, we partition the four elements into two pairs and apply the double-integration method to the difference $[r_{+x}(t) - r_{-x}(t)]$ and $[r_{+y}(t) - r_{-y}(t)]$. This will give the result of the bearing-angle vector $[\theta_x, \theta_y]$ directly. To be more computationally effective, we can apply the double-integration procedure to the weighted version of the differential

$$r_d(t) = (r_{+x}(t) - r_{-x}(t)) + j(r_{+y}(t) - r_{-y}(t))$$

The result of the double-integration procedure produces a complex scalar. The real and imaginary parts of the complex scalar are related to the elements $[\theta_x, \theta_y]$ of the bearing-angle vector, respectively. The computation involved is simple that it can be implemented in the form of first-order op-amp circuits.

To improve the accuracy of the estimation, one approach is to increase the number of receiver elements. For conventional methods, the increase of the number of receiver elements couples with the increase of computation complexity. However, the effect of this technique because of the change of array size is very minor.

The combined differential signal $r_d(t)$ can be regarded as the inner product of the receiver position vector and the received signal vector:

$$\begin{aligned} r_d(t) &= (r_{+x}(t) - r_{-x}(t)) + j(r_{+y}(t) - r_{-y}(t)) \\ &= [1, j, -1, -j][r_{+x}(t), r_{+y}(t), r_{-y}(t), r_{-y}(t)]^T \end{aligned}$$

If we extend this concept to a circular array with N uniformly spaced receiver elements, the sum of the weighted signal is in the form of

$$r_d(t) = \sum_{k=1}^{N} \exp\left(\frac{j2\pi k}{N}\right) r_k(t)$$

The complex weighting coefficients $\{\exp(j2\pi k/N)\}$ are fully defined by the relative positions of the receivers on the two-dimensional plane. This means $r_d(t)$ is a linear combination of the N received signals. After the simple superposition step, the computation procedure is identical with the same demodulation and matched filtering process. This implies that the computation complexity of this algorithm remains largely the same for varying number of receiver elements, which enables significant enhancement of accuracy of bearing-angle estimation without the increase of computation complexity in hardware as well as software.

Traditional algorithms for the estimation of bearing angle are based on the relative time delays among the receivers. This is largely accomplished by cross-correlation procedures, which requires substantial computation. In comparison, the double-integration method is much simpler in terms of computation complexity and hardware structure, with excellent stability and accuracy. By nature, the integration operations of the algorithm provide good tolerance against noise. The simplicity also translates into savings in computation time and power consumption.

Consider the simple case of one transmitter and two receivers. The spacing of the twin receiver pair is D, similar to the previous case. The received signals at the twin receivers are the offset versions of the same signal. We then take the difference of these two signals and apply the double-integration algorithm. From the previous analysis, the final value out of the direct-integration method is in the similar form of

$$\text{Final value} = \frac{cD}{v}|H(0)|^2 \sin(\theta)$$

where θ is the bearing angle with respect to the vector defined by the receiver pair. Now, if we define the angle φ as

$$\varphi = \pi/2 - \theta$$

It can be rewritten in the form of the inner product of the vector $\boldsymbol{u}$ and $\boldsymbol{w}$ as

$$\text{Final value} = \langle \boldsymbol{w}, \boldsymbol{u} \rangle = \frac{cD}{v}|H(0)|^2 \cos(\varphi)$$

where $\boldsymbol{w}$ is the vector defining the three-dimensional bearing-angle vector,

$$\boldsymbol{w} = \frac{c}{v}|H(0)|^2 [\alpha, \beta, \gamma]$$

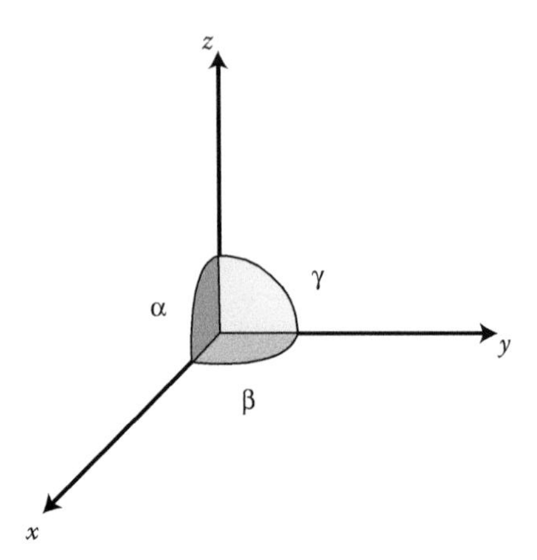

FIGURE 2.14
Directional cosines of the bearing-angle vector.

and α, β, and γ are the directional cosines, $\alpha = \cos(\varphi_x)$, $\beta = \cos(\varphi_y)$, and $\gamma = \cos(\varphi_z)$, respectively, as shown in Figure 2.14.

And $\boldsymbol{u}$ is the vector, defined by the locations of the receiver pair. For example, in the horizontal direction, the vector can be written as $u = D\,[1, 0, 0]$.

Consider the simple case of the circular receiver array with radius $D/2$ and N uniformly spaced receivers. The N receivers are located at the positions of

$$u_n = \frac{D}{2}\exp\left(\frac{j2\pi n}{N}\right)$$

where $n = 0, 1, 2, \ldots, N-1$. Utilizing the circular symmetry, the receivers of the N-element array can be grouped into $N/2$ pairs corresponding to the directional vectors,

$$\boldsymbol{u}_n = D\left[\cos\left(\frac{2\pi n}{N}\right), \sin\left(\frac{2\pi n}{N}\right), 0\right]$$

where $n = 0, 1, 2, \ldots, N/2$. The final value of the integration procedure from the nth pair of receivers is in the form of the inner product of $\boldsymbol{w}$ and $\boldsymbol{u}_n$,

$$\langle \boldsymbol{w}, \boldsymbol{u}_n \rangle = \frac{cD}{v}\,|H(0)|^2\left(\alpha\cos\left(\frac{2\pi n}{N}\right) + \beta\sin\left(\frac{2\pi n}{N}\right)\right)$$

If the received signal is weighted by the location vector of the receivers, the final value of the integration process is weighted by the complex position factor $\exp(j2\pi n/N)$, and it becomes

$$\exp\left(\frac{j2\pi n}{N}\right)\langle w, u_n\rangle = \frac{cD}{v}|H(0)|^2\left[\alpha\cos\left(\frac{2\pi n}{N}\right) + \beta\sin\left(\frac{2\pi n}{N}\right)\right]\exp\left(\frac{j2\pi n}{N}\right)$$

If we can generalize the process by combining the received signal, weighted by the receiver locations, the final value of the integration process is

$$\sum_{n=0}^{N/2-1}\langle w, u_n\rangle \exp\left(\frac{j2\pi n}{N}\right)$$
$$= \sum_{n=0}^{N/2-1}\frac{cD}{v}|H(0)|^2\left(\alpha\cos\left(\frac{2\pi n}{N}\right) + \beta\sin\left(\frac{2\pi n}{N}\right)\right)\exp\left(\frac{j2\pi n}{N}\right)$$

Then, the summation of the first term gives the real part of the complex scalar,

$$\sum_{n=0}^{N/2-1}\alpha\cos\left(\frac{2\pi n}{N}\right)\exp\left(\frac{j2\pi n}{N}\right)$$
$$= \sum_{n=0}^{N/2-1}\alpha\left(\frac{1}{2}\exp\left(\frac{j2\pi n}{N}\right) + \frac{1}{2}\exp\left(\frac{-j2\pi n}{N}\right)\right)\exp\left(\frac{j2\pi n}{2}\right)$$
$$= \sum_{n=0}^{N/2-1}\alpha\left(\frac{1}{2}\exp\left(\frac{j2\pi n}{(N/2)}\right) + \frac{1}{2}\right) = \frac{N}{4}\alpha$$

Similarly, the summation of the second term gives the imaginary component,

$$\sum_{n=0}^{N/2-1}\beta\sin\left(\frac{2\pi n}{N}\right)\exp\left(\frac{j2\pi n}{N}\right) = \sum_{n=0}^{N/2-1}\beta\left(\frac{1}{2j}\exp\left(\frac{j2\pi n}{N}\right)\right)$$
$$-\frac{1}{2j}\exp\left(\frac{-j2\pi n}{N}\right)\exp\left(\frac{j2\pi n}{N}\right) = \sum_{n=0}^{N/2-1}\beta\left(\frac{1}{2j}\exp\left(\frac{j2\pi n}{(N/2)}\right) + \frac{1}{2}j\right) = j\frac{N}{4}\beta$$

The sum of the weighted inner-product terms accumulates to the value of

$$\sum_{n=0}^{N/2-1} \langle w, u_n \rangle \exp\left(\frac{j2\pi n}{N}\right) = |H(0)|^2 \frac{cDN}{4v}(\alpha + j\beta)$$

Figure 2.15 summarizes the estimation procedure of the 8-receiver version of the system. As it can be seen, the main structure of the process, after the weighted-summation process, remains the same for the multi-receiver array. This implies that the computation complexity, in both hardware and software, remains largely the same.

After normalizing the term with the known parameters, the final value of the integration procedure gives the complex scalar $(\alpha + j\beta)$, which represents the first two components of the directional cosines. With the simple relationship, we can obtain the third component of the directional cosines γ,

$$\alpha^2 + \beta^2 + \gamma^2 = 1$$

This analysis can be generalized to perform bearing-angle estimation with a set of N nonuniformly distributed receivers. The steps of the procedure are as follows:

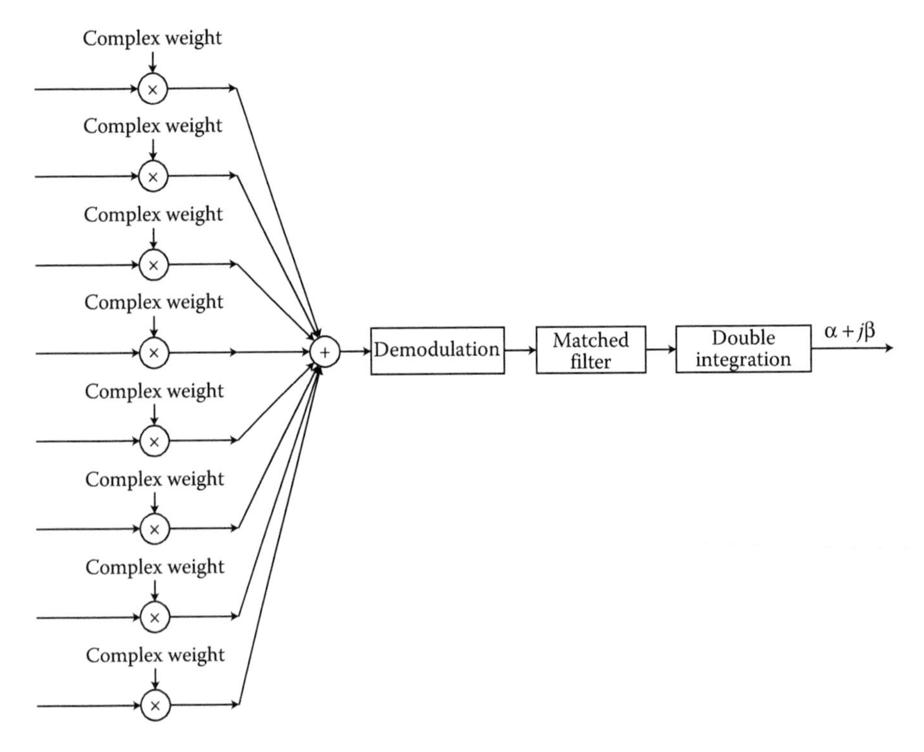

FIGURE 2.15
Procedure of the 8-receiver version of the system.

Step 1: Normalization:

This first step is to perform a simple integration of the received signal at each receiver. This step computes for the average-power term $c|H(0)|^2$ of each channel. Then, the normalization process removes the term subsequently. Normally, this term at all the receiver channels should be the same, since they are same signal with different delays.

Step 2: Complex weighting:

The normalized signals are then weighted with a complex weighting coefficient. Each weighting coefficient contains two components. The phase term is governed by the relative position vector, defined from the mean of all the receiver positions to a particular receiver. The magnitude of this location vector is used to normalize the received waveform. The normalization replaces the $D/2$ factor in the previous discussion. The phase term of the weighting is equivalent to the term $\exp(j2\pi n/N)$.

Step 3: Superposition:

The superposition procedure combines all N tracks of signals from the N receivers into one time-domain signal. Subsequent to the superposition, the combined signal is normalized to remove the factor N and then the factor $2v$ due to propagation.

Step 4: Demodulation and matched filtering:

The combined signal is then demodulated to remove the career component. Subsequently, matched filtering is performed.

Step 5: Integration:

The final step is to integrate the combined signal twice. The result is a complex number $\alpha + j\beta$, of which the real and imaginary parts are the first two directional cosines. The third direction cosine can then be calculated as

$$\gamma = (1 - \alpha^2 - \beta^2)^{1/2}$$

2.5 Active Modality

In this chapter, a simple technique was presented for bearing-angle estimation for UUV underwater geolocation and navigation. The algorithm was originally designed as the replacement for the conventional peak-phase approach for the systems with twin transmission waveforms and the polarity of 180° phase offset. The accuracy, stability, and especially the simplicity of this algorithm made the system, in both hardware and software, significantly more effective.

It turns out that this algorithm can also function effectively with the reversed version of the system. The conventional techniques for passive multi-receiver acoustic arrays were largely structured, based on the computationally extensive cross-correlation method. This simple algorithm showed remarkable system performance in laboratory tests and the simplicity in computation remains for large number of receiver elements.

The existing system is designed for geolocation, navigation, homing and docking operations, with separate transmitter and receiver systems. The receiver component functions as a passive data-acquisition device. Yet, one extremely interesting extension is to modify it into an active system. This can be achieved by placing the transmitter at the center of the circular receiver array.

With the same integration method, this system is capable of accurately estimating the bearing angle of the reflected waveform. Because the transmitted signal is readily available, the time delay between a received waveform $r(t)$ and transmitted signal $T(t)$ can be estimated by using the double-integration method in a similar manner. Subsequently, the range distance of the target can be calculated from the time delay. Details of the range-distance estimation will be given later in Chapter 4. With the capability of dynamic estimation of both the range distance and bearing angle in real time with low power-consumption level, this system can be deployed for collision avoidance for UUV or unmanned aerial vehicle (UAV).

3

System Analysis

In the previous chapter, a simple system for bearing-angle estimation was introduced to illustrate the direct application of signal processing methods. With additional transceivers, the procedure can be extended to the estimation of three-dimensional bearing-angle parameters. It also suggested the extension to the active mode for the inclusion of range estimation.

The key objective of this chapter is to convert the estimation procedure from the acoustical sensor systems for parameter estimation to multi-dimensional imaging. The first step is to model the coherent wave propagation as a linear and shift-invariant system. This model allows us to apply well-established concepts to the signal analysis and processing in imaging system design and development.

As we characterize the data acquisition and image reconstruction as linear systems, it is important to fully formulate the mathematical structure of the process, including the impulse response and transfer function of the systems. This leads to the formulation of the backward-propagation technique as the framework of image formation. In addition, Fresnel and Fraunhofer techniques are introduced as the approximation of the backward-propagation image formation algorithm for the improvement of computation efficiency.

3.1 Linear Coherent Systems

Impulse response and transfer function are the two most common components in system analysis. In this section, these two important elements as well as the relationship are formulated in an organized manner. For simplicity, the impulse functions and plane waves are first analyzed.

3.1.1 Impulse Functions

The impulse function is typically introduced as a time function in signal analysis and processing, in the form of

$$\delta(t) = \begin{cases} \infty & t = 0 \\ 0 & \text{otherwise} \end{cases}$$

with the normalization formula

$$\int \delta(t)\,dt = 1$$

In the field of imaging, the functions include spatial variables, which is in the form of four-dimensional distributions $f(x, y, z; t)$. So, it is necessary to examine the representations and meanings of impulse functions in the space domain.

For the case of three-dimensional systems, the spatial impulse function is in the form of

$$\delta(x,y,z) = \begin{cases} \infty & x = y = z = 0 \\ 0 & \text{otherwise} \end{cases}$$

with unit norm

$$\iiint \delta(x,y,z)\,dx\,dy\,dz = 1$$

This represents *a point source* at the origin. When we move into the two-dimensional case, the impulse function becomes

$$\delta(x,y) = \begin{cases} \infty & x = y = 0 \\ 0 & \text{otherwise} \end{cases}$$

with the two-dimensional norm

$$\iint \delta(x,y)\,dx\,dy = 1$$

Similarly, this represents a point source at the origin of the two-dimensional x–y plane. In addition, in three dimensions, it represents a line source along the z axis.

The following is also a spatial impulse function,

$$\delta(x) = \begin{cases} \infty & x = 0 \\ 0 & \text{otherwise} \end{cases}$$

with unit norm

$$\int \delta(x)\,dx = 1$$

This spatial impulse function has three different representations. In one dimension, it represents a point source at $x = 0$. In two dimensions, it becomes a line source along the y axis over the x–y plane. And, in three dimensions, it represents a plane source, the y–z plane, in the three-dimensional space.

3.1.2 Plane Waves

For simplicity, we first consider a plane wave, with wavelength λ over the two-dimensional x–y plane. As illustrated in Figure 3.1, at the incident angle $\theta = 0°$, the wave pattern observed along the x axis is a constant. In a simple term, the wave pattern can be described as a distribution with spatial frequency $f_x = 0$, and wave pattern can be written as

$$g(x, y) = A$$

where A is the complex amplitude of the observed plane wave.

At the incident angle $\theta = 90°$, as shown in Figure 3.2, the wave pattern observed along the x axis is a single-frequency variation, with the same

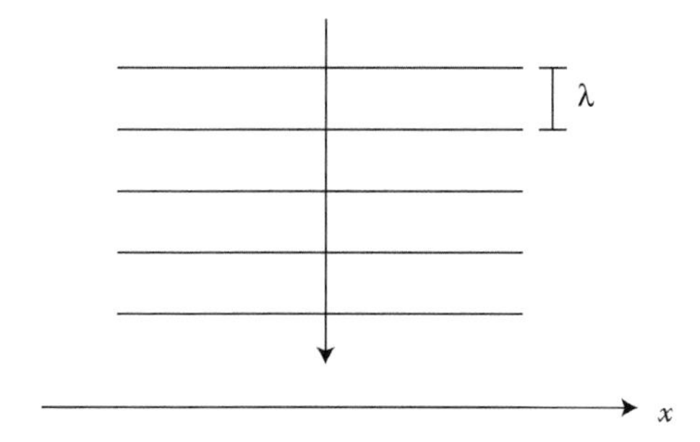

FIGURE 3.1
Plane wave with bearing angle $\theta = 0°$.

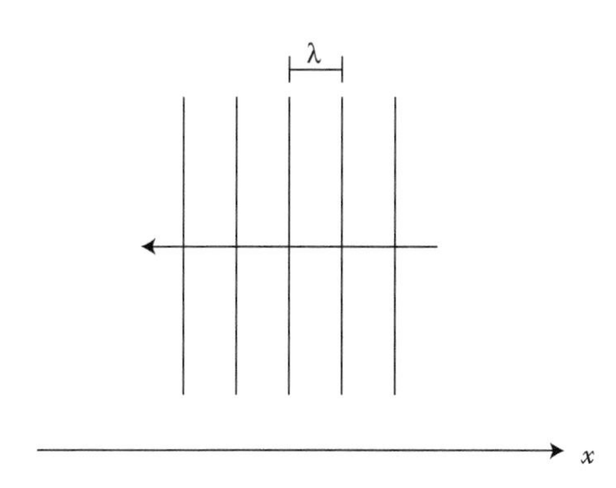

FIGURE 3.2
Plane wave with bearing angle $\theta = 90°$.

wavelength λ. Accordingly, we can represent the wave pattern in the form of

$$g(x,y) = A\exp\left(\frac{j2\pi x}{\lambda}\right)$$

We can also describe it as a coherent pattern with spatial frequency $f_x = 1/\lambda$.

Now we consider the general case that the plane wave is coming in with an arbitrary incident angle θ, as shown in Figure 3.3. Along the x axis, the period of the single-frequency variation is $\lambda/\sin\theta$, corresponding to the spatial frequency $f_x = \sin\theta/\lambda$.

If we also examine the distribution along the y axis, we find the period of the coherent wave distribution is $\lambda/\cos\theta$, corresponding to the spatial frequency $f_y = \cos\theta/\lambda$. Thus, the spatial variation of the plane wave can be written as

$$\begin{aligned} g(x,y) &= A\exp\left(j2\pi\left(\frac{\sin\theta}{\lambda}\right)x\right)\exp\left(j2\pi\left(\frac{\cos\theta}{\lambda}\right)y\right) \\ &= A\exp\left\{j2\pi\left[\left(\frac{\sin\theta}{\lambda}\right)x + \left(\frac{\cos\theta}{\lambda}\right)y\right]\right\} \end{aligned}$$

Then we realize a very interesting relationship

$$f_x^2 + f_y^2 = \left(\frac{\sin\theta}{\lambda}\right)^2 + \left(\frac{\cos\theta}{\lambda}\right)^2 = \frac{1}{\lambda^2}$$

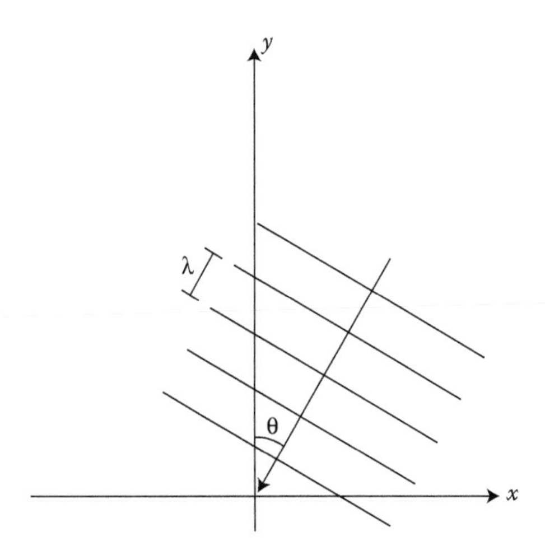

FIGURE 3.3
Plane wave with bearing angle θ.

If we generalize it to the three-dimensional case, the relationship becomes

$$f_x^2 + f_y^2 + f_z^2 = \frac{1}{\lambda^2}$$

In matrix form, we represent the three-dimensional frequency vector as

$$f = [f_x, f_y, f_z]^T$$

Then, the norm of the frequency vector of a plane wave is a constant, defined by the coherent wavelength,

$$\|f\| = \frac{1}{\lambda}$$

The simple property is of great importance. It contains two key elements. One is that the norm of the frequency vector of a plane wave is governed by the wavelength. The other one is that the incident angle of the plane wave is defined by the components of the frequency vector.

3.1.3 Impulse Responses of Wave Propagation

The wavefield produced by a point source has been well established through physics experiments. For three-dimensional cases, the complex phasor version is in the form of

$$h(x, y, z) = \left(\frac{1}{j\lambda r}\right) \exp\left(\frac{j2\pi r}{\lambda}\right)$$

where $r = [x^2 + y^2 + z^2]^{1/2}$, is the distance from the location of the point source to the receiver position In two dimensions, it becomes

$$h(x, y) = \left(\frac{1}{j\lambda r}\right)^{1/2} \exp\left(\frac{j2\pi r}{\lambda}\right)$$

where $r = [x^2 + y^2]^{1/2}$. And for the one-dimensional propagation, it is

$$h(x) = \exp\left(\frac{j2\pi x}{\lambda}\right)$$

It is interesting to point out that these three versions have a similar phase term. The complex-magnitude terms are different, due to the constraints of conservation of power of the wavefield. The wavefield patterns produced by a point source are widely known as *Green's functions*, which is the kernel of the convolution integral. The convolution integral, representing the input–output relationship, is known as the *Rayleigh–Sommerfeld* integral.

As described, wave propagation can be considered as a linear and space-invariant system, with the source distribution as the input and the resultant wavefield as the output. This model allows us to apply Fourier analysis to the design and development of imaging systems. Now, if we regard the point source as an impulse input into the system, the resultant wavefield patterns given here are therefore the *impulse responses* of the system, in different dimensions.

3.1.4 Transfer Functions

Once the impulse response of a linear system is established, the subsequent step is to formulate the transfer function of the system. This allows us to analyze system performance and formulate wavefield scattering in the spatial-frequency domain. In system analysis and signal processing, this is a common approach to the improvement of computation efficiency. Therefore, because of the vital role of the transfer function in imaging system analysis, it is important to formulate the corresponding transfer functions properly.

Normally, the transfer function of a linear system is obtained by Fourier transforming the impulse response directly. However, Fourier transforming Green's functions directly is cumbersome, especially for the two- and three-dimensional cases. So, here an alternative approach is utilized, by moving the source–wavefield relationship back to the basic level of the *Helmholtz equation*,

$$(\nabla^2 + k^2) h(x,y,z) = \delta(x,y,z)$$

where $k = 2\pi/\lambda$. Then, if we consider the waveform outside the source area, which is the case of the imaging applications, the equation can be simplified down to

$$(\nabla^2 + k^2)\, h(x,y,z) = 0$$

where $\nabla^2 = \partial^2/\partial x^2 + \partial^2/\partial z^2 + \partial^2/\partial z^2$. Recalling the frequency-domain equivalence of

$$\partial/\partial x \;\leftrightarrow\; j2\pi f_x$$

we apply Fourier transformation and bring the Helmholtz equation to the spatial-frequency domain, and it becomes

$$\left[-4\pi^2(f_x^2+f_y^2+f_z^2)+\frac{4\pi^2}{\lambda^2}\right]H(f_x,f_y,f_z)=0$$

where f_x, f_y, and f_z are the spatial frequencies in x, y, and z direction, respectively. After simplification, it is now in the form of

$$\left[\frac{1}{\lambda^2}-(f_x^2+f_y^2+f_z^2)\right]H(f_x,f_y,f_z)=0$$

To describe this equation in a simple way, the transfer function $H(f_x, f_y, f_z)$ can be nonzero only when

$$f_x^2+f_y^2+f_z^2=\frac{1}{\lambda^2}$$

If we define the spatial frequency in the vector form, $f=[f_x, f_y, f_z]$, the equation can be written as

$$|f|=\frac{1}{\lambda}$$

This means the transfer function is nonzero only when the magnitude of the spatial-frequency vector is $1/\lambda$. That also means only the wavefield components with the magnitude $|f| = 1/\lambda$ can pass through. Other frequency components will be blocked off by the transfer function. Because of this property, it is logical to formulate the transfer function in the form of

$$H(f_x,f_y,f_z)=\delta\left(|f|-\frac{1}{\lambda}\right)$$

Although the transfer function is organized in the three-dimensional format, this concept remains valid in all dimensions.

3.1.5 Correspondences

The interesting part of this analysis is that the impulse response and transfer function associated with the coherent wave propagation are established independently. As it can be seen, the impulse response is developed through physical experiments, and the transfer function is justified based on mathematical formulation. In order to fully exercise their roles in image reconstruction, the fundamental relationship between the impulse response and transfer function needs to be mathematically structured. Since the impulse response and transfer function are in the form of a Fourier-transform pair, the direct link of Green's functions and transfer functions needs to be fully established. For the three-dimensional case, the inverse transform of the transfer function is

$$\begin{aligned}\mathcal{F}^{-1}\left\{\delta\left(|f|-\frac{1}{\lambda}\right)\right\} &= \left(\frac{2}{\lambda r}\right)\sin\left(\frac{2\pi r}{\lambda}\right) = \left(\frac{1}{j\lambda r}\right)\exp\left(\frac{j2\pi r}{\lambda}\right)+\left(\frac{-1}{j\lambda r}\right)\exp\left(\frac{-j2\pi r}{\lambda}\right)\\ &= h(x,y,z)+h^{*}(x,y,z)\end{aligned}$$

For the two-dimensional case, the mathematical structure gives a similar relationship,

$$\begin{aligned}\mathcal{F}^{-1}\left\{\delta\left(|f|-\frac{1}{\lambda}\right)\right\} &= \left(\frac{2}{\lambda r}\right)J_0\left(\frac{2\pi\lambda}{r}\right) \approx \left(\frac{2}{\lambda r}\right)^{1/2}\cos\left(\frac{2\pi r}{\lambda-\pi/4}\right)\\ &= \left(\frac{1}{j\lambda r}\right)^{1/2}\exp\left(\frac{j2\pi r}{\lambda}\right)+\left(\frac{-1}{j\lambda r}\right)^{1/2}\exp\left(\frac{-j2\pi r}{\lambda}\right)\\ &= h(x,y)+h^{*}(x,y)\end{aligned}$$

And the one-dimensional case is in fact much simpler,

$$\begin{aligned}\mathcal{F}^{-1}\left\{\delta\left(|f|-\frac{1}{\lambda}\right)\right\} &= 2\cos\left(\frac{2\pi x}{\lambda}\right) = \exp\left(\frac{j2\pi x}{\lambda}\right)+\exp\left(\frac{-j2\pi x}{\lambda}\right)\\ &= h(x)+h^{*}(x)\end{aligned}$$

For the three-dimensional case, the transfer function associated with the wave propagation is a sphere in the spatial-frequency domain and the radius of the sphere is $1/\lambda$. In two dimensions, the transfer function is a ring with radius $1/\lambda$. It should also be pointed out that, for all three cases, the transfer functions are real and symmetrical, and so are the corresponding impulse responses. And, Green's function is simply the complex phasor version of the corresponding impulse response.

3.2 Backward Propagation

From the previous section, the transfer function for coherent wave propagation, in three dimensions, is formulated in the form of

$$H(f_x, f_y, f_z) = \delta\left(|f| - \frac{1}{\lambda}\right)$$

Geometrically, this transfer function is a sphere of radius $1/\lambda$. However, the planar receiving aperture is on only one side of the source distribution. In that case, the transfer function does not cover the entire sphere. Instead, we have only half of it, the half toward the receiving aperture. As a result, the transfer function is in the form of

$$H(f_x, f_y, f_z) = \delta\left(|f| - \frac{1}{\lambda}\right) \quad \text{for } f_z \geq 0 = \delta\left(f_z - \left[\frac{1}{\lambda^2} - \left(f_x^2 + f_y^2\right)\right]^{1/2}\right)$$

Then, we perform inverse Fourier transformation of the transfer function, in the z direction only. Utilizing the shift property of the Fourier transform, the transfer function becomes

$$\mathcal{F}_z^{-1}\{H(f_x, f_y, f_z)\} = \mathcal{F}_z^{-1}\left\{\delta\left(f_z - \left[\frac{1}{\lambda^2} - \left(f_x^2 + f_y^2\right)\right]^{1/2}\right)\right\} = H(f_x, f_y; z)$$

$$= \exp(j2\pi f_z z) = \exp\left(j2\pi\left[\frac{1}{\lambda^2} - \left(f_x^2 + f_y^2\right)\right]^{1/2} z\right)$$

This three-dimensional transfer function is a function of one space variable z and two spatial-frequency variables, f_x and f_y. This version of the transfer function plays a profound role in imaging system design and serves as the foundation of the image formation methods. The complete form of the formula is commonly written as

$$H(f_x, f_y; z) = \begin{cases} \exp\left(j2\pi\left[\frac{1}{\lambda^2} - (f_x^2 + f_y^2)\right]^{1/2} z\right) & (f_x^2 + f_y^2) \leq \frac{1}{\lambda^2} \\ 0 & (f_x^2 + f_y^2) > \frac{1}{\lambda^2} \end{cases}$$

This transfer function has several interesting characteristics. Suppose z is a constant, representing the distance from the point source to the planar receiving aperture. If we plot the transfer function over the two-dimensional $f_x - f_y$ plane, the transfer function is lowpass and bounded by a circle of radius $1/\lambda$. It is circularly symmetrical. Within the passband, the transfer function is phase-only with unity magnitude. Figure 3.4 shows the spatial-frequency coverage of the transfer function.

If the wavefield propagates from one plane to another parallel plane, the distance z between the two parallel planes is constant. Thus, the governing transfer function of coherent wave propagation between these two planes is a lowpass filter. The maximum span of the spatial-frequency vector is $1/\lambda$. This also suggests high spatial-frequency components are lost during the propagation.

3.2.1 Image Reconstruction by Inverse Filtering

Conceptually, image reconstruction is to estimate the source distribution with the received wavefield data. This can be regarded as a recovery procedure by reversing the wave scattering process. That is also the reason it is often referred to as *inverse scattering*. The most common inversion procedure is known as inverse filtering, which is simply to apply the inverse of the transfer function as the inverse filter.

Since the plane-to-plane transfer function is lowpass, we can define the inverse filter within the lowpass region, leaving the high-frequency region zero.

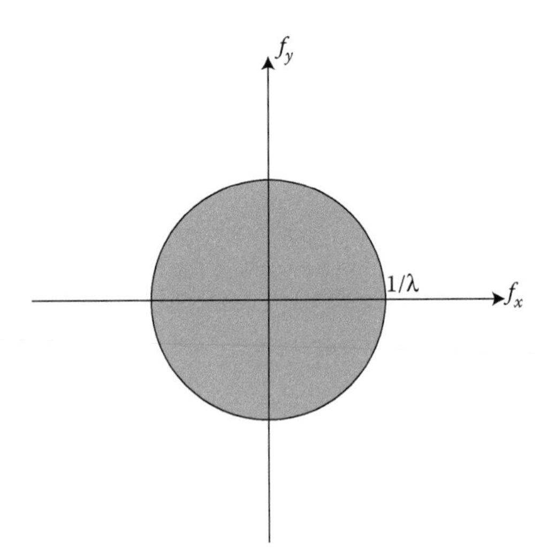

FIGURE 3.4
Spatial-frequency coverage of the transfer function.

$$H^{-1}(f_x, f_y; z) = \exp\left(-j2\pi\left[\frac{1}{\lambda^2} - (f_x^2 + f_y^2)\right]^{1/2} z \right) \qquad \frac{1}{\lambda^2} \geq (f_x^2 + f_y^2)$$

$$0 \qquad \text{otherwise}$$

This also means the reconstruction filter is lowpass with the same cutoff frequency. Then, since the transfer function is phase-only with unity amplitude over the lowpass region, inverting the transfer function is the same as conjugating it,

$$H^{-1}(f_x, f_y; z) = H^*(f_x, f_y; z)$$

This suggests that the inverse filter designed for image reconstruction is in fact the *matched filter* and the image formation process is simply matched filtering.

3.2.2 Plane-to-Plane Backward Propagation

As noted, the image reconstruction filter for the plane-to-plane case involves no more than negating the phase of the transfer function of coherent wave propagation. If we attach the minus sign to the propagation distance z, the filter process associated with image reconstruction is like propagating the received wavefield in the *opposite* direction. It is like treating the received wavefield as a source distribution and propagating it *backward* in the $-z$ direction. Thus, this technique is commonly known as *backward propagation.*

$$H^*(f_x, f_y; z) = \exp\left(-j2\pi\left[\frac{1}{\lambda^2} - (f_x^2 + f_y^2)\right]^{1/2} z \right)$$

$$= H(f_x, f_y; -z) = \exp\left(j2\pi\left[\frac{1}{\lambda^2} - (f_x^2 + f_y^2)\right]^{1/2} (-z) \right)$$

Using the conjugate-symmetry property of the Fourier transform, we can formulate the impulse responses of the backward-propagation algorithm in various dimensions as

$$h^*(x, y, z) = \left(\frac{-1}{j\lambda r}\right) \exp\left(\frac{-j2\pi r}{\lambda}\right)$$

$$h^*(x, y) = \left(\frac{-1}{j\lambda r}\right)^{1/2} \exp\left(\frac{-j2\pi r}{\lambda}\right)$$

$$h^*(x) = \exp\left(\frac{-j2\pi x}{\lambda}\right)$$

This shows the computation structure involved in backward propagation is simplified that both the transfer function and impulse response are the conjugate versions of that of the forward wave propagation process.

3.2.3 Power Spectrum and the Point-Spread Function

Once both the transfer functions of coherent wave propagation and backward-propagation image reconstruction are fully established, we can examine the complete reconstruction process as well as the resolving capability of the imaging process. If we consider the wave propagation as a linear and space-invariant system and the backward-propagation filtering is another linear and space-invariant system, the reconstruction of a centered point source can be described as the result of these two systems in cascade. Since the transfer functions of these two systems are a conjugate pair, the combined system has the transfer function,

$$H(f_x, f_y; z)H^*(f_x, f_y; z) = |H(f_x, f_y; z)|^2$$

$$= 1 \quad \frac{1}{\lambda^2} \geq (f_x^2 + f_y^2)$$

$$0 \quad \text{otherwise}$$

It is important to point out that the combined transfer function represents an ideal lowpass filter, circularly symmetrical, with cutoff radius $1/\lambda$, and unity gain within the passband. The total two-dimensional spectral coverage is

$$\text{Passband area} = \frac{\pi}{\lambda^2}$$

This suggests that higher frequency, corresponding to shorter wavelength, will provide larger spectral bandwidth coverage, which translates into improved resolution.

If we inverse Fourier transform the *power spectrum* back to the space domain, we can obtain the point-spread function,

$$psf(x,y) = \mathcal{F}^{-1}\left\{|H(f_x, f_y; z)|^2\right\}$$

$$= \frac{1}{r} J_1\left(\frac{j2\pi r}{\lambda}\right) \quad r = (x^2 + y^2)^{1/2}$$

where $J_1(j2\pi r/\lambda)$ is the Bessel function. The point-spread function of an imaging system plays a role of great importance. It is the governing element in the assessment of the system's resolving capability.

This point-spread function is well observed. In optics, it is often referred as the *Airy disk.*

Note that this point-spread function is independent of the distance z. This means, for ideal systems with infinite-size apertures, the point-spread function is the same for any range distance z.

3.2.4 Wavefield Signal Processing

We now return to the plane-to-plane case, for simplicity. At plane one, at $z = z_1$, the spatial-frequency spectrum of the wavefield distribution $s(x, y, z_1)$ is in the form of

$$S(f_x, f_y; z_1) = \mathcal{F}_{xy}\{s(x, y, z_1)\}$$

Note that the Fourier transformation takes place in only the x and y direction. Similarly, at plane two, at $z = z_2$, the spatial-frequency spectrum is

$$S(f_x, f_y; z_2) = \mathcal{F}_{xy}\{s(x, y, z_2)\}$$

Based on the linear and space-invariant relationship, the spectra are related through a transfer function,

$$S(f_x, f_y; z_2) = S(f_x, f_y; z_1)\ H[f_x, f_y; (z_2 - z_1)]$$

Assume we detect the wavefield $s(x, y, z_2)$ at $z = z_2$, with an aperture over the x–y plane. We can estimate the wavefield at $z = z_1$, by applying the backward-propagation filter,

$$\begin{aligned} S(f_x, f_y; z_1) &= S(f_x, f_y; z_2)H[f_x, f_y; (z_1 - z_2)] \\ &= S(f_x, f_y; z_2)H^*[f_x, f_y; (z_2 - z_1)] \\ &= S(f_x, f_y; z_2)\exp\left\{-j2\pi\left[\frac{1}{\lambda^2} - (f_x^2 + f_y^2)\right]^{1/2}(z_2 - z_1)\right\} \end{aligned}$$

This operation can also be represented in the space domain in the form of a convolution,

$$s(x, y, z_2) = s(x, y, z_1) * h(x, y, (z_2 - z_1))$$

where Green's function of the integral is

$$h(x,y,(z_2-z_1)) = \left(\frac{1}{j\lambda r}\right)\exp\left(\frac{j2\pi r}{\lambda}\right)$$

where $r = [x^2 + y^2 + (z_2 - z_1)^2]^{1/2}$.

For image reconstruction by backward propagation, the filtering procedure is in the similar form,

$$s(x,y,z_1) = s(x,y,z_2) * h^*(x,y,(z_2-z_1))$$

$$= s(x,y,z_2) * \left(\frac{-1}{j\lambda r}\right)\exp\left(\frac{-j2\pi r}{\lambda}\right)$$

Now, we realize that the backward-propagation integral is very similar to that of the forward propagation. The only difference is that the integration kernel is conjugated. Then, if we conjugate both sides of the equation, it arrives to this interesting form,

$$s^*(x,y,z_1) = s^*(x,y,z_2)^*\left(\frac{1}{j\lambda r}\right)\exp\left(\frac{j2\pi r}{\lambda}\right)$$

This implies that we can reconstruct the conjugated version of the wavefield at $z = z_1$ by forward propagating the conjugated received wavefield detected at $z = z_2$. This means, if we can conjugate the received wavefield physically, we can take advantage of the physical behavior of wave propagation for image reconstruction. As a result, physical wave propagation can replace the computation of image reconstruction. That is also the reason waveform conjugation has been an important research topic in the area of imaging reconstruction.

3.2.5 Backward Propagation in Space Domain

According to the linear space-invariant model, the wavefield generated by a coherent source can be written in the form of a convolution integral over the source region *S*, which is also known as the Rayleigh–Sommerfeld integral in the field of imaging,

$$g(x,y,z) = s(x,y,z) * h(x,y,z)$$

$$= \iiint_S s(x',y',z')h(x-x',y-y',z-z')dx'dy'dz'$$

where $h(x, y, z)$ is Green's function,

$$h(x,y,z) = \left(\frac{1}{j\lambda r}\right)\exp\left(\frac{j2\pi r}{\lambda}\right)$$

and

$$r = (x^2 + y^2 + z^2)^{1/2}$$

The backward-propagation image formation process is also a convolution integral, where the detected wavefield is convolved over the aperture R with the conjugated version of Green's function,

$$\begin{aligned} s'(x,y,z) &= g(x,y,z) * h^*(x,y,z) \\ &= \iiint_R g(x',y',z')h^*(x-x',y-y',z-z')dx'dy'dz' \end{aligned}$$

To visualize and understand the numerical process of image formation, let us use a point source as an example, for simplicity. Subsequently, we can generalize this concept based on the linearity and space invariance.

$$s(x,y,z) = \delta(x - x_0,\ y - y_0,\ z - z_0)$$

The wavefield produced by this point source is in the form of a spatially-shifted Green's function,

$$\begin{aligned} g(x,y,z) &= \delta(x - x_0,\ y - y_0,\ z - z_0) * h(x,y,z) \\ &= h(x - x_0,\ y - y_0,\ z - z_0) \\ &= \left(\frac{1}{j\lambda r}\right)\exp\left(\frac{j2\pi r}{\lambda}\right) \end{aligned}$$

where

$$r = ((x - x_0)^2 + (y - y_0)^2 + (z - z_0)^2)^{1/2}$$

In this case, the backward-propagation image formation is in the form of

$$\begin{aligned} s'(x,y,z) &= g(x,y,z) * h^*(x,y,z) \\ &= \iiint_R \left(\frac{1}{j\lambda r}\right)\exp\left(\frac{j2\pi r}{\lambda}\right)\left(\frac{-1}{j\lambda r'}\right)\exp\left(\frac{-j2\pi r'}{\lambda}\right)dx'dy'dz' \end{aligned}$$

where

$$r = ((x' - x_0)^2 + (y' - y_0)^2 + (z' - z_0)^2)^{1/2}$$

and

$$r' = ((x - x')^2 + (y - y')^2 + (z - z')^2)^{1/2}$$

Then, after reorganizing the terms inside the integral, it becomes

$$\begin{aligned} s'(x,y,z) &= g(x,y,z) * h^*(x,y,z) \\ &= \iiint_R \left(\frac{1}{\lambda^2 r r'}\right) \exp\left(\frac{j2\pi(r - r')}{\lambda}\right) dx'dy'dz' \\ &= \iiint_R A(r,r') \exp(j\theta(r,r')) dx'dy'dz' \end{aligned}$$

In simple words, it can be seen that the image is formed out of an integral of a collection of vectors. The amplitude of the vectors is

$$A(r,r') = \frac{1}{\lambda^2 r r'}$$

and the phase is

$$\exp(j\theta(r,r')) = \exp\left(\frac{j2\pi(r - r')}{\lambda}\right)$$

Thus, the numerical formation of the image at a particular location can be regarded as a sum of a collection of vectors. When, and only when, it is the location of the point source

$$x = x_0, \quad y = y_0, \quad z = z_0$$

The distances r and r' become identical,

$$r = r'$$

As a result, the phase part of the vector becomes zero,

$$\exp(j\theta(r,r')) = \exp\left(\frac{j2\pi(r - r')}{\lambda}\right) = 1$$

Thus, at that location, it becomes an integral of a collection of real and positive values, which produces a large final value

$$s'(x_0, y_0, z_0) = \iiint_R A(r, r')\, dx'dy'dz'$$

$$= \iiint_R \left(\frac{1}{\lambda^2 r'^2}\right) dx'dy'dz'$$

For other locations, on the other hand, it remains to be a vector addition, which induces cancellation due to nonidentical phase terms, producing smaller values. This simple vector accumulation–cancellation effect is the foundation of the image formation process. Larger aperture size improves this vector accumulation–cancellation effect that the final value at the location of the source is greater because of the accumulation of larger number of real and positive values and nonsource area results in with smaller final values due to more effective cancellation. This directly translates into improvement of resolution.

3.2.6 Phase-Only Technique

If we strip off the magnitude of the wavefield data, the integral for image formation becomes

$$s'(x, y, z) = \iiint_R \exp\left(\frac{j2\pi r}{\lambda}\right)\left(\frac{-1}{j\lambda r'}\right)\exp\left(\frac{-j2\pi r'}{\lambda}\right) dx'dy'dz'$$

$$= \iiint_R \left(\frac{-1}{j\lambda r'}\right)\exp\left(\frac{j2\pi(r - r')}{\lambda}\right) dx'dy'dz'$$

After some minor reorganization, we can still represent the integral in the form of vector accumulation–cancellation in a similar manner, except of a complex constant at the front,

$$s'(x, y, z) = \left(\frac{-1}{j}\right)\iiint_R A(r, r')\exp(j\theta(r, r'))\, dx'dy'dz'$$

If so, the amplitude and phase of the vectors are now

$$A(r, r') = \frac{1}{\lambda r'}$$

And

$$\exp(j\theta(r,r')) = \exp\left(\frac{j2\pi(r-r')}{\lambda}\right)$$

respectively. The vector accumulation–cancellation effect remains, which explains the feasibility of image reconstruction using only the phase information of the wavefield.

One very interesting scheme is to further simplify the process by removing the amplitude part of the integration kernel. The phase-only image formation computation becomes an integration of unit-amplitude vectors

$$\begin{aligned} s'(x,y,z) &= \iiint_R \exp\left(\frac{j2\pi r}{\lambda}\right)\exp\left(\frac{j2\pi r'}{\lambda}\right)dx'dy'dz' \\ &= \iiint_R \exp\left(\frac{j2\pi(r-r')}{\lambda}\right)dx'dy'dz' \\ &= \iiint_R \exp(j\theta(r,r'))\,dx'dy'dz' \end{aligned}$$

where

$$\exp(j\theta(r,r')) = \exp\left(\frac{j2\pi(r-r')}{\lambda}\right)$$

This is a special case that the vectors associated with the accumulation–cancellation process are all unit vectors,

$$A(r,r') = A = 1$$

The reason this technique is capable of quality image formation is because the behavior of the vector accumulation–cancellation effect remains exactly the same, although the amplitude portion is removed from the process.

3.3 Fresnel and Fraunhofer Approximations

The objective of the Fresnel and Fraunhofer approximations is to simplify the backward-propagation image formation procedure to the form of Fourier transformation. The approximations are based on the far-field assumption, where the distance between the source plane and receiving aperture is sufficiently large.

3.3.1 Fresnel Approximation

Consider a very simple data-acquisition configuration, where the source plane is located at $z = 0$, and a parallel planar receiving aperture is at $z = z_0$.

$$\begin{aligned} s(x,y,z_0) &= s(x,y,0) * h(x,y,z_0) \\ &= s(x,y,0) * \frac{1}{j\lambda r}\exp\left(\frac{j2\pi r}{\lambda}\right) \\ &= \iint s(x',y',0)h(x-x',y-y',z_0)\,dx'dy' \end{aligned}$$

where $r = [x^2 + y^2 + z_0^2]^{1/2}$. The kernel of the convolution integral is

$$h(x-x',y-y',z_0) = \frac{1}{j\lambda r}\exp\left(\frac{j2\pi r}{\lambda}\right)$$

where r is the distance between a position (x', y') in the source region and the receiving position (x, y),

$$\begin{aligned} r &= \left[(x-x')^2 + (y-y')^2 + (z_0)^2\right]^{1/2} \\ &= z_0\left[1+\left(\frac{x-x'}{z_0}\right)^2 + \left(\frac{y-y'}{z_0}\right)^2\right]^{1/2} \end{aligned}$$

If the propagation distance z_0 is large, we can approximate it by taking the low-order terms of the Taylor's series

$$\begin{aligned} r &\approx z_0\left[1+\frac{1}{2}\left(\frac{x-x'}{z_0}\right)^2 + \frac{1}{2}\left(\frac{y-y'}{z_0}\right)^2\right] \\ &= z_0 + \frac{1}{2z_0}(x^2 - 2xx' + x'^2) + \frac{1}{2z_0}(y^2 - 2yy' + y'^2) \end{aligned}$$

To simplify the procedure, we reorganize them into four terms,

$$r = z_0 + \frac{1}{2z_0}(x^2 + y^2) + \frac{1}{2z_0}(x'^2 + y'^2) - \frac{1}{z_0}(xx' + yy')$$

Then, we place the terms of variable r back into the convolution kernel. Subsequently, it can be partitioned into five terms,

$$h(x-x',y-y',z_0)=\frac{1}{j\lambda r}\exp\left(\frac{j2\pi r}{\lambda}\right)$$

$$\approx\frac{1}{j\lambda z_0}\exp\left(\frac{j2\pi z_0}{\lambda}\right)\exp\left(\frac{j\pi(x^2+y^2)}{\lambda z_0}\right)\exp\left(\frac{j\pi(x'^2+y'^2)}{\lambda z_0}\right)\exp\left(\frac{-j2\pi(xx'+yy')}{\lambda z_0}\right)$$

We should note that the complex amplitude term $1/j\lambda r$ is approximated as $1/j\lambda z_0$. This approximation is feasible, because amplitude of the integration kernel is less sensitive to approximation errors.

Now, we plug the five-term approximation back into the original plane-to-plane propagation model, and it becomes

$$s(x,y,z_0)=s(x,y,0)*h(x,y,z_0)$$

$$=\iint s(x',y',0)h(x-x',y-y',z_0)dx'dy'$$

$$\approx\frac{1}{j\lambda z_0}\exp\left(\frac{j2\pi z_0}{\lambda}\right)\exp\left(\frac{j\pi(x^2+y^2)}{\lambda z_0}\right)$$

$$\iint s(x',y',0)\exp\left(\frac{j\pi(x'^2+y'^2)}{\lambda z_0}\right)\exp\left(\frac{-j2\pi(xx'+yy')}{\lambda z_0}\right)dx'dy'$$

The first two terms are complex constants governed by the constant distance z_0 and we can move them out of the integral. The third term, $\exp(j\pi(x^2+y^2)/\lambda z_0)$, is a phase term, which is not a function of the integration variables, x' and y'. So, we can also move it outside the integral.

Here, it should be pointed out that the quadratic phase term $\exp(j\pi(x^2+y^2)/\lambda z_0)$ is commonly known as the *Fresnel phase pattern*, which is circularly symmetrical. This phase factor has important applications in various imaging tasks.

Now, we group the three terms outside the integral as

$$A(x,y)=\frac{1}{j\lambda z_0}\exp\left(\frac{j2\pi z_0}{\lambda}\right)\exp\left(\frac{j\pi(x^2+y^2)}{\lambda z_0}\right)$$

Then, we define a new term $s'(x, y, 0)$ as

$$s'(x,y,0)=s(x,y,0)\exp\left(\frac{j\pi(x^2+y^2)}{\lambda z_0}\right)$$

Mathematically, this term is the source distribution $s(x, y, 0)$ multiplied by the Fresnel phase mask $\exp(j\pi(x^2 + y^2)/\lambda z_0)$. Now, the *Rayleigh–Sommerfeld* integral is simplified down to

$$\begin{aligned} s(x,y,z_0) &= A(x,y)\iint s'(x',y',0)\exp\left(\frac{-j2\pi(xx'+yy')}{\lambda z_0}\right)dx'dy' \\ &= A(x,y)\iint s'(x',y',0)\exp\left(-j2\pi\left(\frac{x}{\lambda z_0}\right)x'\right)\exp\left(-j2\pi\left(\frac{y}{\lambda z_0}\right)y'\right)dx'dy' \\ &= A(x,y)S'\left(\frac{x}{\lambda z_0},\frac{y}{\lambda z_0},0\right) \end{aligned}$$

where $S'(f_x, f_y, 0)$ is the two-dimensional Fourier transform of the modulated source distribution $s'(x', y', 0)$.

This relationship is one of the most important establishments in coherent imaging. It means the resultant wavefield $s(x, y, z_0)$ over a planar aperture at $z = z_0$ is related to a scaled version of the Fourier transform of the masked source distribution $s'(x', y', 0)$ at $z = 0$. Therefore, the coherent wave propagation from the source plane $z = 0$ to the receiving plane $z = z_0$ can be summarized by the steps:

1. Multiply the source distribution $s(x, y, 0)$ by the Fresnel phase mask $\exp(j\pi(x^2 + y^2)/\lambda z_0)$.
2. Fourier transform the masked source distribution $s'(x, y, 0)$ to obtain $S'(f_x, f_y, 0)$.
3. Scale the spectrum pattern $S'(f_x, f_y, 0)$ by the scaling factor $1/\lambda z_0$ to produce the pattern $S'(x/\lambda z_0, y/\lambda z_0, 0)$.
4. Multiply by the term $(1/j\lambda z_0)\exp(j2\pi z_0/\lambda)\exp(j\pi(x^2 + y^2)/\lambda z_0)$ to obtain $s(x, y, z_0)$.

The important feature of this approach is that the convolution integral can be replaced by a Fourier-transformation operation, with a scaling and multiplication process. In addition, we should note that all the steps of the procedure are reversible. This means that we can perform image reconstruction from the wavefield pattern over the receiving aperture through Fourier transformation with the use of Fresnel phase masks. Figure 3.5 shows the mathematical model of wave propagation through Fresnel approximation.

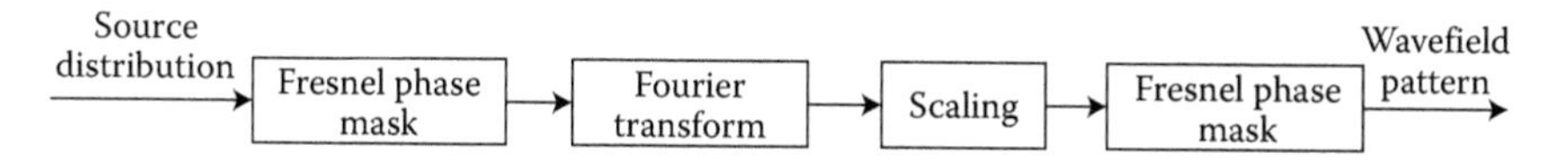

FIGURE 3.5
Wave propagation model through Fresnel approximation.

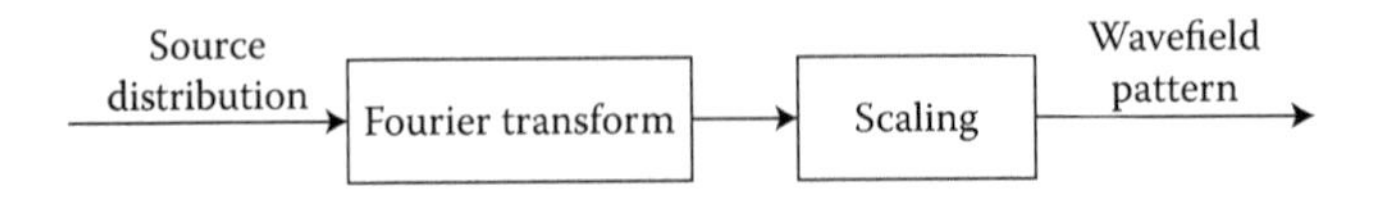

FIGURE 3.6
Wave propagation in the form of Fraunhofer approximation.

3.3.2 Fraunhofer Approximation

If the source region is significantly small with respect to the range distance, the Fresnel phase factor is no longer effective.

$$\exp\left(\frac{j\pi(x^2+y^2)}{\lambda z_0}\right) \approx 1$$

Thus, the term $A(x, y)$ can also be approximated as a complex scalar,

$$A(x,y) = \frac{1}{j\lambda z_0}\exp\left(\frac{j2\pi z_0}{\lambda}\right)\exp\left(\frac{j\pi(x^2+y^2)}{\lambda z_0}\right) \approx \frac{1}{j\lambda z_0}\exp\left(\frac{j2\pi z_0}{\lambda}\right) = A$$

As a result, the Fresnel approximation procedure can be further approximated as

$$s(x,y,z_0) \approx A(x,y)\iint s(x',y',0)\exp\left(\frac{-j2\pi(xx'+yy')}{\lambda z_0}\right)dx'dy' = A\ S\left(\frac{x}{\lambda z_0},\frac{y}{\lambda z_0},0\right)$$

Then, first step, applying the phase mask, can also be eliminated. So, it is now simplified down to a Fourier-transformation operation. Thus, the source distribution and received wavefield pattern are related as a Fourier-transform pair, which is known as the *Fraunhofer approximation*. This important relationship has been widely utilized in far-field applications, especially in the area of *synthetic-aperture radar* (*SAR*) *imaging*. Figure 3.6 shows the mathematical model of wave propagation in the form of Fraunhofer approximation.

3.4 Holographic Imaging

The objective of this section is, through the presentation of the classical holography, to illustrate the importance of the use of reference waves for data acquisition and the role of the phase information for image reconstruction.

Illuminating from the object, the complex wavefield observed at the viewing position is in the form of

$$\text{Object beam} = [U_o(x,y)\,\exp(j\phi(x,y))]\,\exp(-j\omega t)$$

where $U_o(x,y)$ and $\exp(j\phi(x,y))$ are the magnitude and phase component of the wavefield, and $\exp(-j\omega t)$ is the temporal carrier of the propagating acoustic waves. Also arriving at the viewing position, the reference waveform, which is normally a plane wave, is

$$\text{Reference beam} = [U_r \exp(j2\pi f_{xr} x)]\exp(-j\omega t)$$

where U_r is the complex amplitude of the plane wave and f_{xr} is the spatial frequency,

$$f_{xr} = \frac{\sin\theta}{\lambda}$$

Thus, at the recording position, the complete waveform, which is the superposition of the object beam and reference beam, becomes

$$\begin{aligned} U(x,y) &= \text{Object beam} + \text{Reference beam} \\ &= [U_o(x,y)\exp(j\phi(x,y)) + U_r \exp(j2\pi f_{xr} x)]\exp(-j\omega t) \end{aligned}$$

Then, the intensity of the waveform consists of the components of

$$\begin{aligned} I(x,y) &= |U(x,y)|^2 = U(x,y)U*(x,y) \\ &= \left[U_o^2(x,y) + |U_r|^2\right] \\ &\quad + [U_r * U_o(x,y)\,\exp(j\phi(x,y)\exp(-j2\pi f_{xr} x)] \\ &\quad + [U_r U_o(x,y)\,\exp(-j\phi(x,y)\exp(j2\pi f_{xr} x)] \end{aligned}$$

The recording of the waveform is operating in the linear region of receivers. As a result, the transparency function of the holographic plate can be written as

$$T(x,y) = kI(x,y)$$

During the holographic reconstruction process, the object is removed and a reconstruction beam is utilized. For simplicity, the reconstruction beam is arranged at the same position with the identical configuration. Behind the holographic plate, the observed waveform is the transparency function modulated by the reconstruction beam,

$$\begin{aligned} U_{\text{out}}(x,y) &= T(x,y)\left[U_r \exp(j2\pi f_{xr}x)\right]\exp(-j\omega t) \\ &= kI(x,y)[U_r \exp(j2\pi f_{xr}x)]\exp(-j\omega t) \\ &= k\left[U_o^2(x,y)+|U_r|^2\right][U_r\exp(j2\pi f_{xr}x)]\exp(-j\omega t) \\ &\quad + k[|U_r|^2 U_o(x,y)\exp(j\phi(x,y)]\exp(-j\omega t) \\ &\quad + k[U_r^2 U_o(x,y)\exp(-j\phi(x,y)\exp(j2\pi(2f_{xr})x)]\exp(-j\omega t) \end{aligned}$$

As it can be seen, there are three components in the reconstructed waveform. The first term is a real and positive profile $k\,[U_o^2(x,y) + |U_r|^2]$ modulated by the reconstruction beam $U_r\exp(j2\pi f_{xr}\,x)\exp(-j\omega t)$,

$$\text{1st term} = k\left[U_o^2(x,y)+|U_r|^2\right]\left[U_r \exp(j2\pi f_{xr}x)\right]\exp(-j\omega t)$$

The range information of the object distribution is largely retained in the phase variation. Thus, this term does not contribute much information to the formation of the object profile. Physically, this term is observed in the form of a plane wave with a slightly blurred magnitude distribution.

The second term is the most crucial component. It is exactly the object waveform scaled by a constant $k|U_r|^2$,

$$\begin{aligned} \text{2nd term} &= k\left[|U_r|^2 U_o(x,y)\exp(j\phi(x,y)\right]\exp(-j\omega t) \\ &= k|U_r|^2[U_o(x,y)\exp(j\phi(x,y)]\exp(-j\omega t) \end{aligned}$$

Thus, this component gives the three-dimensional profile, exactly the same as if we were observing the object from the viewing position.

The third is also interesting. It is the complex conjugate of the object waveform, scaled by a complex constant kU_r^2,

$$\begin{aligned} \text{3rd term} &= k\left[U_r^2 U_o^*(x,y)\exp(-j\phi(x,y))\exp(j2\pi(2f_{xr})x)\right]\exp(-j\omega t) \\ &= (kU_r^2)[U_o^*(x,y)\exp(-j\phi(x,y))\exp(j2\pi(2f_{xr})x)]\exp(-j\omega t) \end{aligned}$$

This term is coming to the viewer from a different incident angle φ, corresponding to the spatial frequency $2f_{xr}$,

$$\frac{\sin\varphi}{\lambda} = 2 f_{xr} = 2\frac{\sin\theta}{\lambda}$$

The relationship between the incident angle φ and the incident angle of the reconstruction wave is in the form of

$$\varphi = \sin^{-1}(2\sin\theta)$$

One interesting observation is the similarity between holography and the modulation and demodulation of signals. The reference wave is equivalent to the modulation signal in amplitude modulation (AM) procedure, and the reconstruction wave is equivalent to the demodulation signal. The spatial frequency, f_{xr}, of the reference and reconstruction waves is equivalent to the modulation frequency in AM modulation and demodulation.

The key difference is that the holographic recording is not capable of executing the multiplication process in AM modulation. To facilitate the operation, it utilizes the addition operation in the form of wavefield superposition. With the addition of a reference wave, the recording the intensity allows us to retain the phase information. This process also produces the first and third terms as the byproducts of the process.

3.5 Diffraction Tomography

The formulation of coherent wavefield is governed by the Helmholtz equation that the spectral content of the coherent wavefield is concentrated over a sphere of radius 1/λ, where λ is the operating wavelength. From the perspective of a linear system, the propagation can be described in the form of a transfer function,

$$H(f) = \delta\left(|f| - 1/\lambda\right)$$

where $f = (f_x, f_y, f_z)$ is the spatial-frequency vector and f_x, f_y, and f_z are spatial frequencies in the x, y, and z directions, respectively. In this section, derivations are given in two dimensions for simplicity. In two dimensions, the spectral distribution of the resultant coherent wavefield can be illustrated as a ring of radius 1/λ.

In the ideal case of data acquisition with an infinite linear receiving aperture, the detected acoustic wavefield is a lowpass signal with the cutoff spatial frequency at the maximum span of 2/λ, from –1/λ to 1/λ. By backward propagating the received wavefield to the source region, we reconstruct the wavefield distribution corresponding to a spectral content of the semicircle toward the direction of the aperture span. The other half of the circle is not available in the spectrum of the received wavefield, because the receiving aperture is on one side of the source only.

3.5.1 Active Illumination

In diffraction tomography, the imaging operation normally involves active illumination. Mathematically, we describe the received wavefield as the result of the *secondary reactive radiation* caused by the illumination waveform.

Consider the following imaging scheme that the object distribution $g(x, y)$ is illuminated by a coherent plane wave of wavelength λ and the spatial-frequency vector $f = (f_x, f_y) = (0, -1/\lambda)$ defining the direction of the active illumination from the bottom. Mathematically, the wavefield generated by the illumination plane wave can be written as

$$s(x,y) = \exp\left(\frac{-j2\pi y}{\lambda}\right)$$

The object is then modulated by the illumination wavefield and becomes a secondary source, and the resultant wavefield is then detected at a linear aperture at the top, assumed to be of infinite size at this point, with the aperture span along $y = y_0$. The detected wavefield along the linear aperture is termed a *projection*, and denoted as $p(x)$. This configuration is typically referred to as the transmission imaging mode where the receiver array is on the opposite side of the transmitter.

As indicated previously, the projection $p(x)$ is a lowpass signal with cutoff frequency of $\pm 1/\lambda$. By backward propagating the projection $p(x)$ to the object region, an image subcomponent is then formed, and the spectral distribution of this image component is the lower half of the circle, the semicircle toward the direction of the aperture. The other half of the circular spectrum corresponds to the reflection-mode wavefield, which can be obtained by placing an aperture on the same side of the transmitter. We should also note that the transmission and reflection modes share different portions of the spectral distribution of the resultant wavefield. Figure 3.7 shows the spectral distribution of the received wavefield of the transmission and reflection modes.

Because the resultant wavefield is caused by the scattering due to the illumination instead of direct radiation from the object $g(x, y)$, the semicircular spectrum is associated with a secondary source function instead of the object distribution itself. Mathematically, the secondary source function can be described by $g'(x, y)$ as the object distribution modulated by the wavefield of the plane-wave illumination.

$$g'(x,y) = g(x,y)s(x,y) = g(x,y)\exp\left(\frac{-j2\pi y}{\lambda}\right).$$

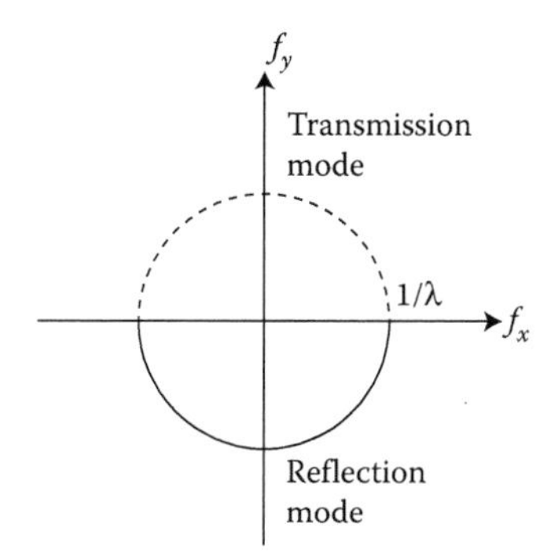

FIGURE 3.7
Spectral distribution of the received wavefield of transmission and reflection modes.

To perform image formation, we first need to properly relate the backward propagated image component to the object distribution, by first removing the contribution from the illumination wavefield. Because of the use of plane wave for illumination, the demodulation procedure is relatively straightforward. It can be performed in the space domain, or even more effectively in the spatial-frequency domain by shifting the spectral distribution backward by $1/\lambda$ toward the direction of the transmitter. As a result, the spectrum of the demodulated backward propagated image component is an offset semicircle. Figure 3.8 shows the spectral distribution of the demodulated wavefield of the transmission and reflection modes, respectively.

The semicircle spectral distribution corresponding to the transmission-mode image component *always* passes through the origin, independent of the operating wavelength. As the operating frequency of illumination signal increases, the operating wavelength becomes shorter, the span of the semicircle becomes larger, and the curvature of the circle decreases

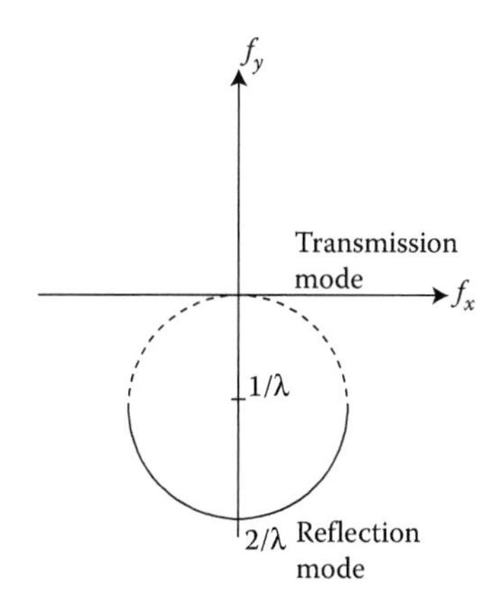

FIGURE 3.8
Spectral distribution of the demodulated wavefield of transmission and reflection modes.

accordingly. When the operating frequency of the diffraction tomographic system approaches infinity, the spectral content of a backward propagated image component from one single projection, especially in the low spatial-frequency region, approaches a linear segment passing through the origin. This means, in terms of the spectral content as well as image formation, the transmission-mode diffraction tomography is approaching the classical case of x-ray tomography as the operating frequency approaches infinity. In addition, from a different perspective, we can describe the conventional x-ray tomography as a limiting case of the generalized transmission-mode diffraction tomography, and the spectrum analysis presented here in this paper can be viewed as the generalized version of the *Central Projection Theorem.*

3.5.2 Tomographic Reconstruction

As it is illustrated in the previous section, the spectral coverage of the image component reconstructed from one single projection is quite limited, and so is the resolving capability, consequently. In order to improve the resolution of the system, one approach is to perform a rotational scan to generate additional coverage of the spectral content, which can be achieved by rotating either the data-acquisition system or object specimen.

Because Fourier transformation is rotation invariant, as the object rotates, the spectral distribution of the object function also rotates accordingly. Therefore, the rotational scan enables us to collect a sequence of semicircular spectral components from various angles and form a circular spectral patch through tomographic superposition. For transmission-mode systems, the resultant spectral coverage produced by the rotational tomographic scan is a circular disk and the radius is $\sqrt{2}/\lambda$. For the reflection mode, the final spectral coverage is a circular band, and the upper and lower bounds of the band are $2/\lambda$ and $\sqrt{2}/\lambda$, respectively (Figure 3.9).

It is important to point out that the transmission-mode imaging produces a spectral coverage completely different from that generated by reflection-mode

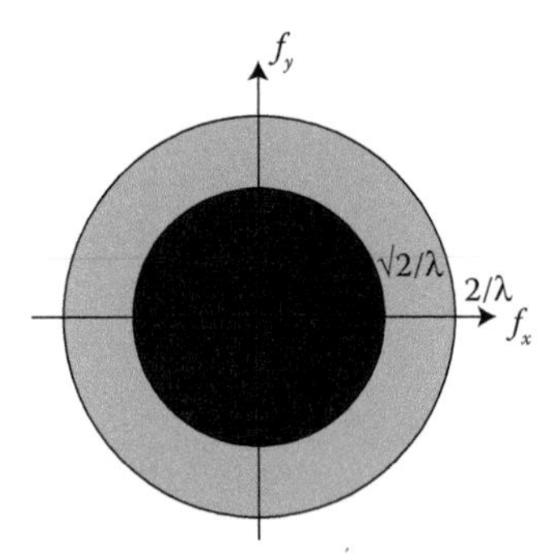

FIGURE 3.9
Spectral coverage of the transmission-mode and reflection-mode images after full rotational scan.

operations. These two spectral areas are mutually orthogonal without any overlapping. The spatial-frequency band of the transmission-mode tomographic images lies in the lowpass region, while the reflection mode covers the bandpass domain. Respectively, the point-spread functions of the transmission- and reflection-mode systems are

$$h_{\text{tr}}(r) = \left(\frac{1}{r}\right) J_1\left(\frac{2\sqrt{2}\pi r}{\lambda}\right)$$

$$h_{\text{ref}}(r) = \left(\frac{1}{r}\right) J_1\left(\frac{4\pi r}{\lambda}\right) - \left(\frac{1}{r}\right) J_1\left(\frac{2\sqrt{2}\pi r}{\lambda}\right)$$

where J_1 denotes the first-order Bessel function.

It is also interesting to note that both the transmission and reflection modes cover exactly the same amount bandwidth area in the spatial-frequency domain.

$$\text{Passband area} = \frac{2\pi}{\lambda^2}$$

This means, even though the point-spread functions of these two operating modes are different in nature, the levels of information content are identical, and therefore, these two modes have the same level of resolving capability.

3.5.3 Finite-Size Aperture

The analysis so far has been limited to the case of infinite receiving aperture. However, the size of the receiving aperture is normally finite in practice. When the size of the receiving aperture becomes limited, the detected wavefield does not extend to the full-span of the spectral coverage as in the cases of infinite-size aperture. The cutoff frequency of the lowpass wavefield is consequently reduced from the upper limit of $1/\lambda$ to

$$f_c = \left(\frac{1}{\lambda}\right) \sin\left(\frac{\theta}{2}\right)$$

where the term $\sin(\theta/2)$ is a factor given by the *Rayleigh criteria* as a function of aperture size and range distance. When it is projected into the two-dimensional space in the spatial-frequency domain, the spectrum covers only an arc segment instead of a semicircle. The span of the arc is directly related to the size of the receiving aperture, as given in the equation. According to the formulation of the lowpass cutoff frequency in the resolution analysis, the angular span of the arc is exactly θ. This means the angular span of the arc is the same as angular coverage defined by aperture with respect to the source region. It is also important to point out that the angular span of the wavefield spectrum is governed by the aperture size and is independent of the operating mode, transmission or reflection, of the imaging system. Hence, the angular span is identical for both the transmission and receiving modes when the aperture size is fixed. After the demodulation process to remove the contribution due to the illumination wavefield, the spatial-frequency spectral content shifts similarly.

A tomographic angular scan is then conducted in a similar manner to expand the spectral coverage. As a result, the complete spectral distribution of the transmission mode is now a disk of reduced area, and the reflection-mode coverage becomes an annular band with reduced width. Quantitatively, the radius of the circular patch corresponding to the transmission-mode operation is

$$R = \left(\frac{2}{\lambda}\right)\sin\left(\frac{\theta}{4}\right)$$

and the upper and lower bounds of the annular band of the reflection-mode spectrum are

$$R_{\text{max}} = \frac{2}{\lambda}$$

$$R_{\text{min}} = \left(\frac{2}{\lambda}\right)\cos\left(\frac{\theta}{4}\right)$$

respectively. Once again, the transmission and reflection modes, with limited aperture size, produce different spectral coverage without any overlapping area. The transmission mode covers the low-frequency domain while the reflection mode covers the bandpass region. It is also important to note these two spectral distributions have identical spatial-frequency coverage area of

$$\text{Spectral coverage} = \left(\frac{2\pi}{\lambda^2}\right)\left(1 - \cos\left(\frac{\theta}{2}\right)\right)$$

and thus have the equivalent level of resolving capability. With a finite-size receiving aperture, the point-spread functions of the transmission- and reflection-mode tomographic reconstruction become

$$h_{\text{tr}}(r) = \left(\frac{R_{\text{min}}}{r}\right)J_1(2\pi R_{\text{min}}r)$$

$$h_{\text{ref}}(r) = \left(\frac{R_{\text{max}}}{r}\right)J_1(2\pi R_{\text{max}}r) - \left(\frac{R_{\text{min}}}{r}\right)J_1(2\pi R_{\text{min}}r)$$

It can be easily seen that, if the angular coverage approaches the full 180° span, the spectrum reaches the maximum content corresponding to the case of infinite-size aperture, as we expect. The illustration of spectral coverage has been concentrated mainly on the transmission or reflection modes in this paper. With slight modifications, we can generalize the analysis into various data-acquisition configurations, mono-static or bi-static.

One can easily visualize that the spectral content of an arbitrary data-acquisition scheme with coherent illumination will be an annulus. The radial maximum and minimum of the annulus are $2/\lambda$ and zero, respectively. The cross-sectional width of the annulus, equivalently the variation of the upper and lower bounds, is governed by the angular coverage defined by the aperture as well as the angular offset between the direction of illumination and the location of the receiving aperture. This implies that, for each single-frequency coherent illumination, the tomographic scan produces a well-defined coverage in the spatial-frequency domain corresponding to a given aperture size. By extending the illumination signal to stepped-frequency FMCW (frequency-modulated continuous wave) waveforms, we can then effectively expand the spectral coverage to achieve desired resolving capability.

Yet, the most intriguing development from this analysis is when the imaging system configuration approaches the other extreme where the data acquisition is performed with a single transmitter receiver in mono-static operating mode. In this case, the angular span of the spectral content, θ, becomes zero, which becomes a limiting case of the reflection mode. As a result, the spectral coverage is reduced from a circular band to a ring of radius $2/\lambda$. This result is in complete agreement with that derived from the coherent SAR operating in spotlight mode.

In SAR imaging, the resolving capability is improved by expanding the spectral coverage from a ring to an annulus with wideband illumination such as the stepped chirp waveforms. And, as illustrated in this paper, a similar expansion of spectral coverage can be achieved with a designated aperture size. This gives the equivalence and a direct trade-off relationship between the temporal bandwidth of illumination signals and spatial span of receiving apertures. This relationship allows us further to the optimization of system resolution by combining the consideration of aperture configurations with signaling schemes.

3.6 Resolution Analysis of Discrete Arrays

The resolving capability of an imaging system is the most critical parameter for system performance evaluation. Over the years, the most widely applied formula for resolution analysis has been largely based on the *Rayleigh*

Resolution Criteria. This simple yet very powerful formula was established in the optics area, based on observations in the laboratories. The resolution limit is defined as the smallest separation between two point sources, still separable by the imaging system. In this section, alternatively, we try to establish the resolution limit of an imaging system from the perspective of Fourier analysis. This allows us to perform system design and evaluation in a clear and quantitative manner.

3.6.1 Classical Formulation

For one-dimensional case, according to *Rayleigh resolution criteria*, the formula for cross-range resolution of a centered point source by a passive coherent imaging system is in the form of

$$\Delta x = \frac{0.5\,\lambda}{\sin(\theta/2)}$$

where λ is the coherent wavelength, and θ is the angular span defined by the aperture with respect to the location of the point source,

$$\sin(\theta/2) = \left(\frac{(R/2)}{(R/2)^2 + z^2}\right)^{1/2}$$

and R is the size of the aperture and z is the distance from the source to the aperture. When range distance z is sufficiently large, it can be further simplified as

$$\sin(\theta/2) \approx \tan(\theta/2) = \frac{R}{2z}$$

and the Rayleigh resolution limit is in the simple form of

$$\Delta x = \frac{0.5\,\lambda}{\sin(\theta/2)} \approx \lambda\left(\frac{z}{R}\right)$$

This is often referred to as the *small-aperture case*.

It implies that the resolution of a system is governed mainly by three parameters. The first is the wavelength, and longer wavelength translates into poorer resolving capability. So, for coherent systems, higher frequencies, corresponding to shorter wavelengths, improves the resolution. The second is the range distance, and farther distance degrades resolution. Then, the third is the aperture size, and larger receiver aperture gives better resolution, which is often demonstrated by the utilization of large lenses in optical systems.

This formula can be easily expanded to the case of two-dimensional rectangular apertures, and the two-dimensional resolution cell has the size of

$$\Delta x = \frac{0.5\,\lambda}{\sin(\theta/2)}$$

and

$$\Delta y = \frac{0.5\,\lambda}{\sin(\psi/2)}$$

where θ and ψ are the angular spans, defined by the aperture, in the x and y direction, respectively.

The case of two-dimensional circular apertures is most interesting. For a circular aperture with diameter R, the Rayleigh resolution limit is the same in all directions on the x–y plane because of the symmetry of the imaging device,

$$\Delta s = \Delta x = \Delta y = \frac{0.61\lambda}{\sin(\theta/2)} \cong \frac{1.22\,\lambda}{\sin(\theta)} \cong 1.22\lambda\frac{z}{R}$$

The unique feature is the extra constant of 1.22, which will be examined mathematically later in this chapter.

3.6.2 Fourier Analysis

For simplicity, we first examine the resolving capability of a passive coherent imaging system. Resulted from a centered point source, the coherent wavefield over the aperture is in the form of

$$\left(\frac{1}{j\lambda r}\right)\exp\left(\frac{j2\pi r}{\lambda}\right) = \left(\frac{1}{j\lambda r}\right)\exp\left(\frac{j2\pi(z^2+x^2)^{1/2}}{\lambda}\right) \cong \left(\frac{1}{j\lambda z}\right)\exp\left(\frac{j2\pi(z^2+x^2)^{1/2}}{\lambda}\right)$$

where $r = (z^2 + x^2)^{1/2}$ is the distance from the aperture region to the target position, and the linear aperture is located along the cross-range direction x. By definition, the local spatial frequency of the wavefield at a position x over the aperture is

$$f_x = \frac{\partial}{\partial x}\left[\frac{(z^2+x^2)^{1/2}}{\lambda}\right] = x/\lambda\,(z^2+x^2)^{1/2}$$

$$= \left(\frac{1}{\lambda}\right)\left[\frac{x}{(z^2+x^2)^{1/2}}\right] = \left(\frac{1}{\lambda}\right)\sin(\varphi)$$

where φ is defined as

$$\sin(\varphi) = \frac{x}{(z^2 + x^2)^{1/2}}$$

over the interval of $(-\theta/2, \theta/2)$. This means that the spatial-frequency bandwidth over the aperture region is a function of the perspective angle φ, with the maximum wing span from $(1/\lambda)\sin(-\theta/2)$ to $(1/\lambda)\sin(\theta/2)$. Thus, the spatial-frequency bandwidth for a symmetric aperture can be approximated as

$$\Delta f_x = f_{\max} - f_{\min} = \left(\frac{1}{\lambda}\right)\sin\left(\frac{\theta}{2}\right) - \left(\frac{-1}{\lambda}\right)\sin\left(\frac{\theta}{2}\right) = \left(\frac{2}{\lambda}\right)\sin\left(\frac{\theta}{2}\right)$$

Thus the cross-range resolution for a passive coherent imaging system is

$$\Delta x = \lambda/2 \sin\left(\frac{\theta}{2}\right)$$

This analysis, based on spatial-frequency bandwidth, matches the experimental observations widely known as the *Rayleigh resolution limit*. It should also be noted here that, for the extreme case of infinite-size aperture with the full angular span of $\theta = \pi$, the resolution reaches the limit of half wavelength,

$$\Delta x = \frac{\lambda}{2}$$

Thus, half wavelength has long been regarded as the fundamental limit of passive coherent systems. Because, in this case, the spatial-frequency band covers the interval of $(-1/\lambda, +1/\lambda)$ with the maximum bandwidth of $2/\lambda$, half wavelength has also been commonly used as the sample spacing of receiver arrays, corresponding to the *Nyquist rate*.

If the aperture becomes circular covering an angular span of θ, the two-dimensional spatial-frequency band is a circular pupil with diameter of

$$\Delta f_{xy} = \Delta f_s = \left(\frac{2}{\lambda}\right)\sin\left(\frac{\theta}{2}\right)$$

and a two-dimensional spatial–spectral coverage area of

$$\Delta S = \pi\left(\frac{1}{\lambda}\right)^2 \sin^2\left(\frac{\theta}{2}\right)$$

If we inverse Fourier transform the two-dimensional circular pupil spectrum,

$$S(f_x, f_y) = 1 \quad \text{for } (f_x^2 + f_y^2)^{1/2} \leq \left(\frac{1}{\lambda}\right)\sin\left(\frac{\theta}{2}\right)$$

the result in the space domain is the Airy disk with the first zero crossing at

$$\Delta s = \frac{0.61\,\lambda}{\sin(\theta/2)} \approx \frac{1.22\,\lambda}{\sin(\theta)} \cong \frac{1.22\,z\lambda}{R}$$

where R is now the diameter of the circular aperture. The constant of 1.22 came as a scaling factor from the Fourier–Bessel transform, due to the circular symmetry.

3.6.3 Active Systems

For active multi-static systems, the target is illuminated by the transmitted waveforms and the reflected wavefield is then detected over the aperture. From a transmitter located at x_0, the target area is modulated by the illumination waveform of

$$m_+(x) = \left(\frac{1}{j\lambda r}\right)\exp\left(\frac{j2\pi r}{\lambda}\right) = \left(\frac{1}{j\lambda r}\right)\exp\left(\frac{j2\pi(z^2 + (x - x_0)^2)^{1/2}}{\lambda}\right)$$

$$\cong \left(\frac{1}{j\lambda z}\right)\exp\left(\frac{j2\pi(z^2 + (x - x_0)^2)^{1/2}}{\lambda}\right)$$

The waveform can be regarded as a modulation pattern, and the local modulation frequency is

$$f_x = \frac{\partial}{\partial x}\left[\frac{(z^2 + (x - x_0)^2)^{1/2}}{\lambda}\right] = \frac{(x - x_0)}{\lambda(z^2 + (x - x_0)^2)^{1/2}} = \left(\frac{1}{\lambda}\right)\left[\frac{(x - x_0)}{(z^2 + (x - x_0)^2)^{1/2}}\right]$$

And at the neighborhood of the point source, $x = 0$, the spatial frequency of the modulation waveform is

$$\left(\frac{1}{\lambda}\right)\left[\frac{-x_0}{\left(z^2 + x_0^2\right)}\right]^{1/2} = \sin\frac{(-\varphi)}{\lambda} = \frac{-\sin(\varphi)}{\lambda}$$

So, locally, the modulation waveform can be approximated as

$$m_+(x) \cong \left(\frac{1}{j\lambda z}\right)\exp\left(\frac{-j2\pi x\,\sin(\varphi)}{\lambda}\right)$$

Physically, $m_+(x)$ is equivalent to a plane wave with wavelength λ and incident angle $-\varphi$. This modulation process shifts the frequency band of the received wavefield from the original spectral interval of $\{-(1/\lambda)\sin(\theta/2), +(1/\lambda)\sin(\theta/2)\}$ down to the frequency interval of $\{-(1/\lambda)\sin(\theta/2) - (1/\lambda)\sin(\varphi), (1/\lambda)\sin(\theta/2) - (1/\lambda)\sin(\varphi)\}$. Similarly, from the transmitter at the opposite end, the modulation waveform is

$$m_-(x) \cong \left(\frac{1}{j\lambda z}\right)\exp\left(\frac{+j2\pi x\sin(\varphi)}{\lambda}\right)$$

This modulation waveform shifts the local spatial-frequency band from the original interval of $\{-(1/\lambda)\sin(\theta/2), (1/\lambda)\sin(\theta/2)\}$ up to $\{-(1/\lambda)\sin(\theta/2) + (1/\lambda)\sin(\varphi), (1/\lambda)\sin(\theta/2) + (1/\lambda)\sin(\varphi)\}$.

Then, we consider the full-span of the aperture. The upper and lower bounds of the modulation frequencies are corresponding to the transmitters at the far opposite ends, $\varphi = \pm\theta/2$. Combining these two components, the composite spatial–spectral interval becomes $\{-(2/\lambda)\sin(\theta/2), (2/\lambda)\sin(\theta/2)\}$, which translates to the spatial bandwidth of

$$\begin{aligned}\Delta f_x = f_{\max} - f_{\min} &= \left(\frac{2}{\lambda}\right)\sin\left(\frac{\theta}{2}\right) - \left(\frac{-2}{\lambda}\right)\sin\left(\frac{\theta}{2}\right)\\ &= \left(\frac{4}{\lambda}\right)\sin\left(\frac{\theta}{2}\right)\end{aligned}$$

doubling the original spectral coverage. This improves the cross-range resolution by *a factor of two* for using the active multi-static illumination. Therefore, the resolution in the cross-range direction is

$$\Delta x = \left(\frac{\lambda}{4}\right)\sin\left(\frac{\theta}{2}\right)$$

For the extreme case of infinite-size aperture with the full angular span of $\theta = \pi$, the resolution reaches the limit of quarter wavelength,

$$\Delta x = \frac{\lambda}{4}$$

And for multi-frequency multi-static systems, the maximum spatial-frequency bandwidth in the cross-range direction is

$$\Delta f_x = \left(\frac{4}{\lambda_{\min}}\right)\sin\left(\frac{\theta}{2}\right) = \frac{4\sin(\theta/2)}{\lambda_{\min}}$$

where $\lambda_{\min}$ is the shortest wavelength of the multi-frequency illumination, corresponding to the highest frequency, and θ denotes the maximum angular span defined by the aperture, with respect to the source position x. This means the resolution in the cross-range direction is

$$\Delta x_{\min} = \frac{\lambda_{\min}}{4\sin(\theta/2)}$$

If the angle θ is small, it can be approximated further as

$$\Delta x_{\min} = \frac{\lambda_{\min}}{2\sin(\theta)}$$

3.6.4 Range Resolution

The classical resolution analysis for coherent systems did not cover the range resolution. Historically, there were two main reasons for the lack of analysis. One is that the classical resolution analysis came from the optics area, where the capability of separating point sources was demonstrated with a screen at fixed distance, parallel to the aperture. The other reason is that the resolution in the range direction was known to be poor for optical devices such as lenses. So, there was no sufficient interest.

Because the mathematical framework for resolution analysis in terms of spatial-frequency bandwidth is fully established here, we can easily extend the analysis to cover the resolution in the range direction. Again, the coherent wavefield over the aperture is in the form of

$$m_+(x) = \left(\frac{1}{j\lambda r}\right)\exp\left(\frac{j2\pi r}{\lambda}\right) = \left(\frac{1}{j\lambda r}\right)\exp\left(\frac{j2\pi(z^2+(x-x_0)^2)^{1/2}}{\lambda}\right)$$

$$\cong \left(\frac{1}{j\lambda z}\right)\exp\left(\frac{j2\pi(z^2+(x-x_0)^2)^{1/2}}{\lambda}\right)$$

The spatial frequency of the wavefield in the z direction is

$$f_z = \frac{\partial}{\partial z}\left[\left(z^2+(x-x_0)^2\right)^{1/2}/\lambda\right] = \frac{z}{\lambda}(z^2+(x-x_0)^2)^{1/2}$$

In the neighborhood of $x = 0$, the modulation frequency becomes

$$f_z = \left(\frac{1}{\lambda}\right)\left[\frac{z}{\left(z^2 + x_0^2\right)^{1/2}}\right] = \frac{1}{\lambda}\cos(\varphi)$$

Again, the spatial-frequency bandwidth over the aperture region is a function of the perspective angle φ, with the maximum range from $(1/\lambda)$ to $(1/\lambda)$ $\cos(\theta/2)$. Accordingly, the spatial-frequency bandwidth in the z direction becomes

$$\Delta f_z = \left(\frac{1}{\lambda}\right)\left[1 - \cos\left(\frac{\theta}{2}\right)\right]$$

Thus, the range resolution for coherent systems is then in the form of

$$\Delta z = \frac{\lambda}{(1 - \cos(\theta/2))}$$

For the extreme case of infinite-size aperture with the full angular span of $\theta = \pi$, the resolution reaches the limit of

$$\Delta z = \lambda$$

On the other hand, when the aperture is small and $\cos(\theta/2) \approx 1$, the resolution in the range direction is very poor, corresponding to the extremely narrow bandwidth Δf_z. That is why wideband waveforms are commonly applied to improve the spatial-frequency bandwidth and consequently the range resolution.

Also in this section, a sequence of diagrams is provided to graphically illustrate the distributions of the spatial-frequency spectra of various array configurations. For simplicity, the diagrams are prepared in two dimensions.

Figure 3.10a shows the simple arrangement of one radiating source and one receiver. In terms of modality, this is considered a passive imaging mode. In the frequency domain, the spectral content provided by the receiver is one sample along the circle of radius $1/\lambda$, with exactly the same perspective angle θ_o.

Similarly, for a passive system with three receivers, the spectrum consists of three frequency samples along the circle of radius $1/\lambda$, with exactly the same perspective angles, as shown in Figure 3.11.

If the system is operating in the active mono-static mode, the frequency sample will be shifted to the circle of radius $2/\lambda$, as shown in Figure 3.12. The relocation of the frequency content is due to the modulation and modulation effects of the wavefield.

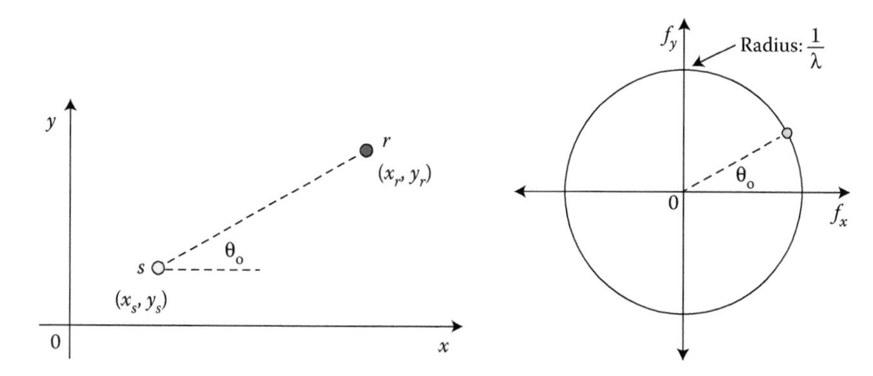

FIGURE 3.10
Source–receiver geometry and spectral content.

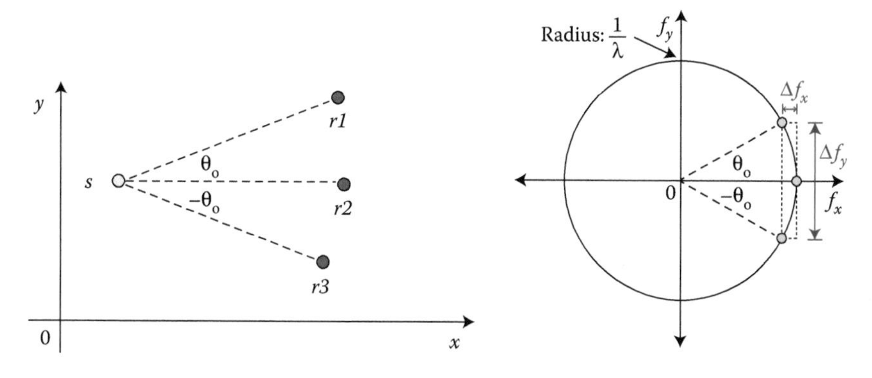

FIGURE 3.11
Source–receiver geometry and spectral content.

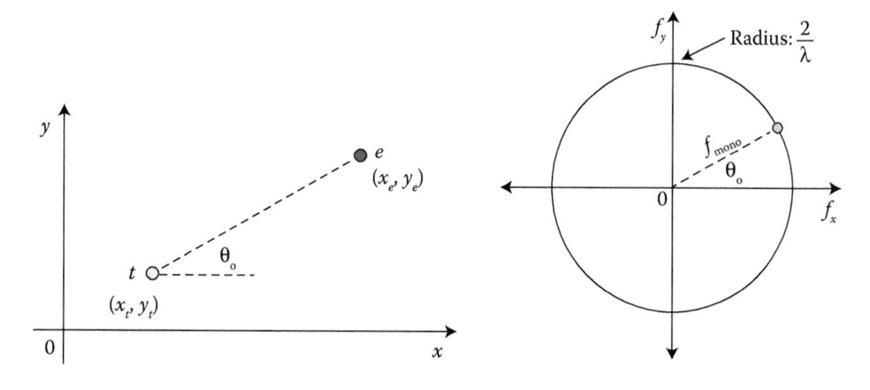

FIGURE 3.12
Mono-static transmitter–receiver geometry and spectral content.

Shown in Figure 3.13, the active bi-static mode involves the sum of two vectors, which moves the spectral sample to a location, bounded by the circles of radius $1/\lambda$ and $2/\lambda$.

By extending this concept, we can visualize the distribution of the spectral contents of the multi-static imaging modality. As shown in Figure 3.14, the multi-static system's array consists of five transceivers. Figure 3.15a is the spectral sample distribution provided by five transceiver elements operating in the multi-static mode with one operating frequency, and Figure 3.15b shows the distribution for three frequencies.

3.6.5 Wideband Case

If we consider the case of wideband waveforms instead of coherent signals, with a temporal bandwidth B,

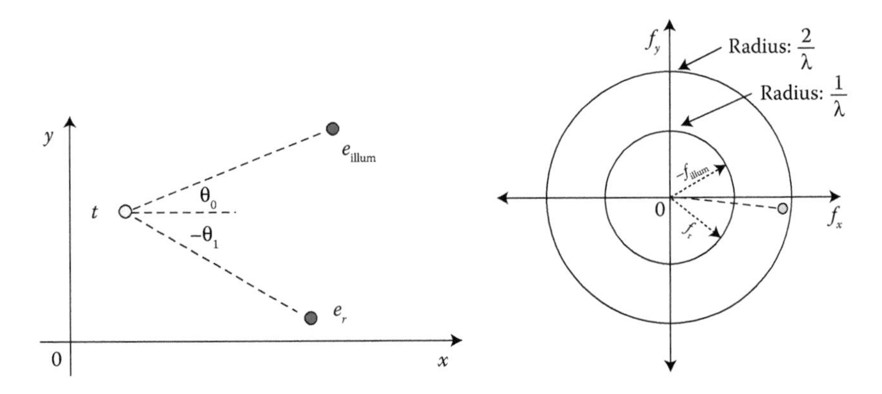

FIGURE 3.13
Bi-static transmitter–receiver geometry and spectral content.

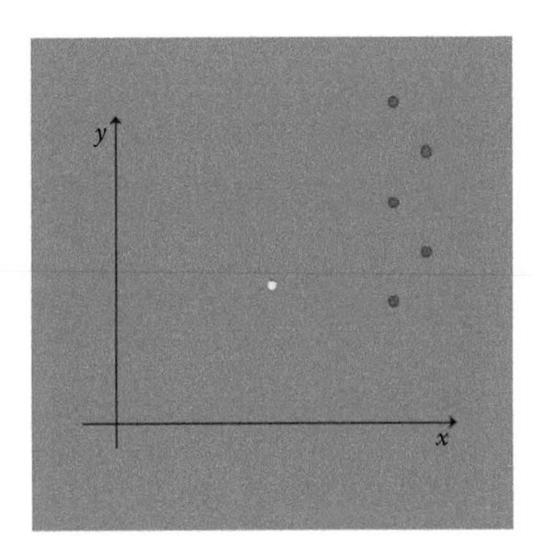

FIGURE 3.14
Multi-static imaging system with an array of five transceivers.

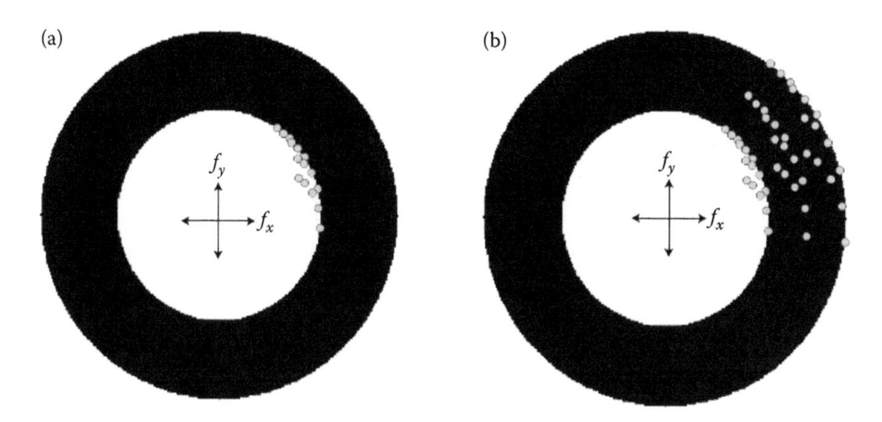

FIGURE 3.15
(a) Spectral distribution with one operating frequency and (b) three frequencies.

$$\Delta f_t = B = f_2 - f_1$$

where f_1 and f_2 are the lower and upper bounds of the temporal frequency band. Their corresponding wavelengths are

$$\lambda_{max} = \frac{v}{f_1}$$

and

$$\lambda_{min} = \frac{v}{f_2}$$

For the wavefield component corresponding to the wavelength λ_{max}, spatial-frequency band covers from $(1/\lambda_{max})$ to $(1/\lambda_{max})\cos(\theta/2)$, and $(1/\lambda_{min})$ to $(1/\lambda_{min})\cos(\theta/2)$ for the wavelength λ_{min} and operating frequency f_2. In combination, the overall spatial-frequency band becomes

$$\Delta f_z = f_{max} - f_{min} = \left(\frac{1}{\lambda_{min}}\right) - \left(\frac{1}{\lambda_{max}}\right)\cos\left(\frac{\theta}{2}\right)$$

$$= \left(\frac{1}{\lambda_{min}}\right) - \left(\frac{1}{\lambda_{max}}\right) + \left(\frac{1}{\lambda_{max}}\right)\left[1 - \cos\left(\frac{\theta}{2}\right)\right]$$

$$= \frac{(f_2 - f_1)}{v} + \left(\frac{f_1}{v}\right)\left[1 - \cos\left(\frac{\theta}{2}\right)\right] = \left(\frac{1}{v}\right)\left\{B + f_1\left(1 - \cos\left(\frac{\theta}{2}\right)\right)\right\}$$

This spatial-frequency bandwidth translates into the range resolution in the space domain,

$$\Delta z = \frac{v}{[B + f_1(1 - \cos(\theta/2))]}$$

This simple formula gives the range resolution in terms of both the waveform bandwidth and aperture coverage, which is of great interest and importance to the design of advanced imaging systems.

In the case of small aperture and large bandwidth, it reduces to

$$\Delta z = \frac{v}{B}$$

If the wideband signals are utilized for the active imaging systems as the illumination waveforms, the range resolution improves further by a factor of two, due to the round-trip propagation,

$$\Delta z = \frac{v}{2B}$$

We can now summarize the resolution analysis in a simple manner:

In the category of coherent passive systems, the governing parameters are (i) the range distance z, (ii) aperture size R, and (iii) wavelength λ. Shorter wavelength, larger aperture size, and smaller range distant improve the resolution.

For active systems, the defining parameters remain the same with similar governing relationships. The resolution improves by a factor of two due to the modulation effect.

For wideband systems, the governing parameters for cross-range resolution are similar, including the range distance z, aperture size R, and *shortest* wavelength λ_{min} within the bandwidth, corresponding to the highest frequency.

In the range direction, the range resolution is governed by (i) the temporal bandwidth B and (ii) propagation speed v. Wider bandwidth and slower propagation speed improve the range resolution.

It is important to note that the resolution analysis is performed based on the assumption of a centered point target, for simplicity. If the target moves away from the center with an offset, the formula needs to be modified accordingly. This also implies that the resolving capability over the target region is not exactly uniform. The center of the target region, with respect to the orientation of the aperture, normally is the position for optimal resolution.

4

Acoustical Imaging Applications

Chapter 3 provided the foundation of the linear shift-invariant model of wave propagation. The model is now fully established with the impulse response, transfer function, and resolution analysis. This allows Fourier analysis to take the center role of system modeling and image reconstruction.

The analysis in Chapter 3 was based on the concept of coherent wave propagation, which is the building block of the linear model. The objective of this chapter is to extend from the coherent model to the wideband format, in order to formulate the resolution analysis in the range direction for practical applications. For the use of wideband waveforms as the probing signals, this chapter starts with the pulse–echo modality, and then transitions into the use of chirp signal and step-frequency FMCW (frequency-modulated continuous wave) waveforms for illumination. Acoustic microscopy, synthetic-aperture sonar imaging, and step-frequency medical ultrasound are used as examples to demonstrate the concepts and capability of these techniques.

As formulated in the image formation process, the procedure consists of two components. One is conducted over the aperture region, and the other is over the spatial-frequency spectrum. Because of linearity, the order of these two steps can be reversed. The reversed version performs the estimation of the range profiles first. As a result, once the range profiles are computed, the organized configuration of the transceiver aperture is no longer significant in the image formation procedure. This opens the door to the important concept of reconfigurable arrays for dynamic imaging.

4.1 Multi-Frequency Imaging

The coherent backward propagated image is in the form of

$$\begin{aligned}\hat{s}(x,y,z;\lambda) &= g(x,y,z;\lambda) * h^*(x,y,z;\lambda)\\ &= \iiint_R g(x',y',z';\lambda)\, h^*(x-x',y-y',z-z';\lambda)\, dx'\, dy'\, dz'\end{aligned}$$

where $g(x,y,z;\lambda)$ denotes the coherent wavefield, corresponding to the single wavelength λ, detected at the receiver aperture. This is equivalent to holographic imaging, illustrated in Chapter 3. The difference is that the phase information can be obtained directly from the received waveforms by the use of quadrature receivers. One direct application of holographic imaging is the acoustic microscopy, when it is operated in the coherent mode.

4.1.1 Scanning Tomographic Acoustic Microscopy

Conventional acoustic microscopy has been limited to the imaging of thin planar specimens. The objective of the scanning tomographic acoustic microscopy (STAM) is to achieve three-dimensional acoustic imaging at the microscopic scale and advance the resolving capability especially in the depth direction. STAM's imaging modality is single-frequency illumination in the transmission mode with multiple observation angles. The operating frequency of the illumination acoustic plane waves is 100 MHz, and the data acquisition of the acoustic wavefield is performed by a focused scanning laser beam, and subsequently through a knife-edge detector. Figure 4.1 shows the STAM prototype system, converted from a conventional scanning laser acoustic microscope (SLAM).

Subsequent to the knife-edge detector, the signal is down-converted from 100 to 32.4 MHz, followed by a quadrature receiver. This allows the detection of both the amplitude and phase of the waveforms.

When it is operated in the holographic imaging mode, the STAM is an excellent example of the plane-to-plane backward propagation image formation. Because the STAM system is capable of the detection of the complex wavefield, it allows us to reconstruct three-dimensional images with complex amplitude. For the use of coherent plane waves, the phase information of

FIGURE 4.1
Scanning tomographic acoustic microscope.

the image can be converted into the propagation time delay, which is directly related to the propagation velocity of the specimens. Since the propagation velocity of material is directly related to the density, the phase information can be mapped into the density distribution of the specimen. This technique is applied to the visualization of the hardening of biological tissues. Figure 4.2a is the interference pattern of a piece of liver specimen, which is the conventional approach to the estimation of tissue hardening. Figure 4.2b is the density profile computed from the phase profile of the STAM image. It shows clearly the hardening areas of the specimen.

Because of the availability of the phase information, the detected waveforms can be backward propagated down toward the three-dimensional subsurface region to form a holographic image. A range finder is used to determine and demonstrate the focusing and resolving capability of subsurface imaging. Figure 4.3a is the image of a subsurface range finder from the conventional intensity-mapping SLAM system and Figure 4.3b is the STAM image of the subsurface layer, illustrating the focused image of the range finder.

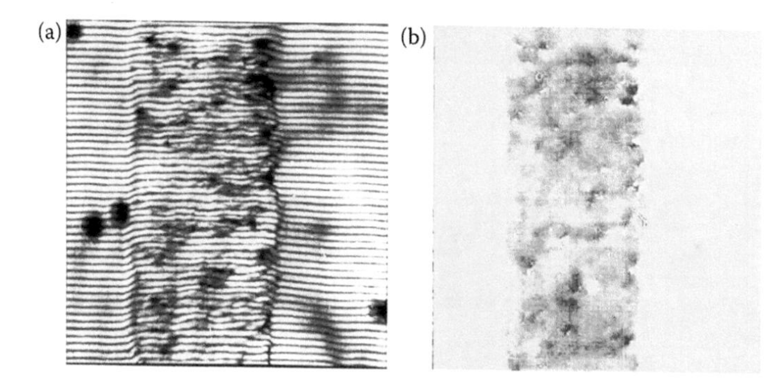

FIGURE 4.2
(a) Interference pattern of the liver specimen and (b) density distribution converted from phase profile the complex STAM image.

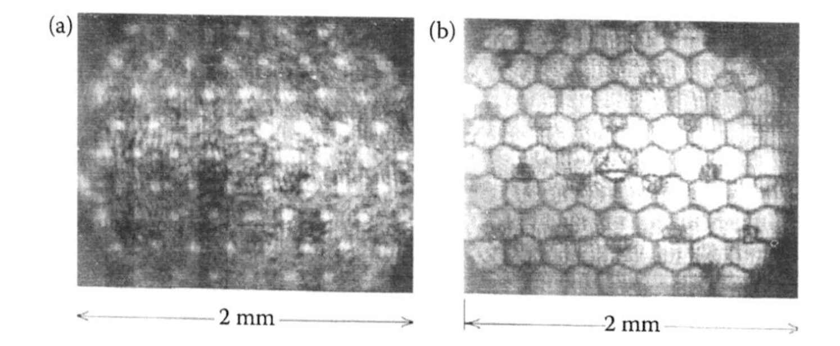

FIGURE 4.3
(a) SLAM image of a range finder. (b) STAM image.

4.1.2 Multi-Frequency Backward Propagation

If we repeat the data-acquisition process by varying the wavelength, λ becomes a variable. And the final image is the superposition of all the coherent sub-images

$$\hat{s}(x,y,z) = \int_{\Lambda} \hat{s}(x,y,z;\lambda)\, d\lambda$$

$$= \int_{\Lambda}\iiint_{R} g(x',y',z';\lambda)\, h^{*}(x-x',\, y-y',\, z-z';\lambda)\, dx'\, dy'\, dz'\, d\lambda$$

where Λ denoted the span of the wavelength variation. The four-dimensional integration can be partitioned into two steps. One is the integration over the aperture region R, and the other is the superposition over the wavelength span Λ. Figure 4.4 shows the general structure of the backward propagation image reconstruction algorithm operating in the multi-frequency modality.

Tomographic acoustic imaging can be enhanced with the multiple-frequency data. This is to operate the data-acquisition process with a

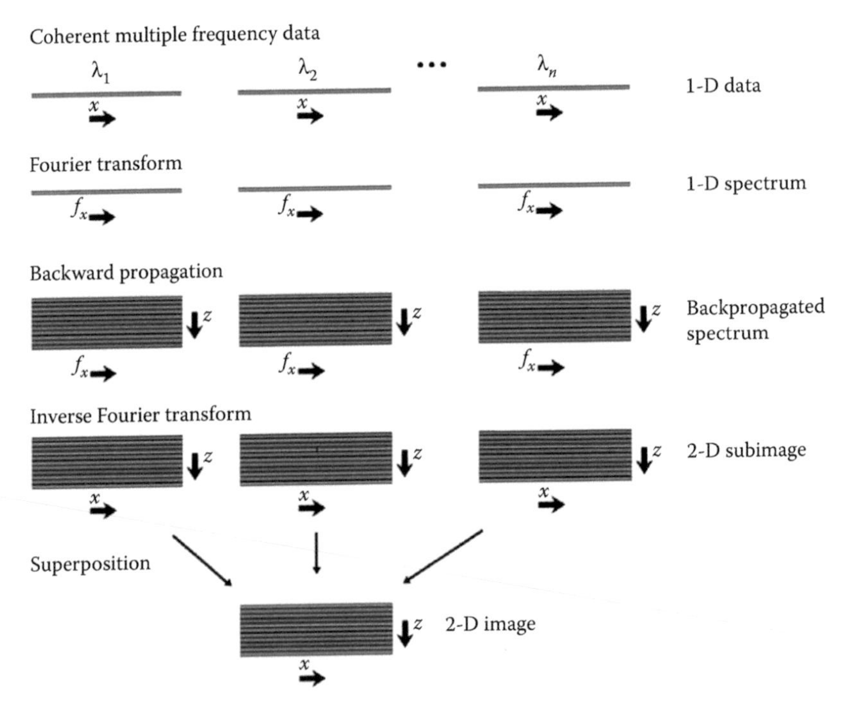

FIGURE 4.4
Structure of the backward propagation image reconstruction algorithm operating in the multi-frequency modality.

collection of continuous wave (CW) frequencies. It enables us to further improve the resolution in the range direction. Figure 4.5 shows the experiment of the imaging of a penny. The transmission-mode waveforms are detected on the opposite side of the penny by the STAM system. Then the detected complex waveforms are backward propagated to form the image of the surface profile of the penny at the opposite side.

These two integration steps of the backward propagation algorithm are independent and the order can be reversed,

$$\begin{aligned}\hat{s}(x,y,z) &= \iiint_R \int_\Lambda g(x',y',z';\lambda)h^*(x-x',y-y',z-z';\lambda)\,d\lambda\,dx'\,dy'\,dz' \\ &= \iiint_R \hat{s}(x-x',y-y',z-z')\,dx'\,dy'\,dz'\end{aligned}$$

Of great interest is the one-dimensional integration

$$\hat{s}(x-x',y-y',z-z') = \int_\Lambda g(x',y',z';\lambda)\,h^*(x-x',y-y',z-z';\lambda)\,d\lambda$$

It should be pointed that the $\hat{s}(x-x', y-y', z-z')$ is a sub-image over the source region over (x, y, z) with the range profile formulated at the receiver position at (x', y', z').

From the sequential arrangement of the image formation procedure, the reconstruction process can be implemented in two versions.

The first version is to conduct coherent backward propagation to form a collection of coherent sub-images. And the final image is the superposition of all the sub-images. The computation is devoted to mainly the backward

FIGURE 4.5
STAM image of the penny formed from the waveforms at the opposite side.

propagation procedure for the formation of coherent sub-images. This technique is effective for system with structured aperture configurations, such as linear and circular arrays, because the backward propagation reconstruction procedure can be modeled and implemented as a convolution process.

The second approach is to form a collection of range profiles. Then the final image is the superposition of the collection of sub-images, which are formed from the range profiles. The main computation is allocated to the estimation of the range profiles. This approach does not require organized structure of the aperture arrays. Therefore, it is suitable to systems with nonuniform or dynamically reconfigurable arrays. On the other hand, the frequency spacing needs to be uniform to achieve effective computation for the range profiles.

These two versions produce the same final image. For mono-static imaging modality, these two have the same level of computation complexity. For systems operating in bi-static or multi-static modes, the second version is consistently more effective and efficient.

4.2 Pulse–Echo Imaging

The pulse–echo model is the most traditional imaging configuration, which is also the conceptual foundation of many imaging reconstruction techniques. In this section, we present the structure of the image formation algorithm of the FMCW multi-static imaging systems from the perspective of the traditional pulse–echo approach, for simplicity.

The pulse–echo technique is considered a reflection-mode space–time-domain approach, commonly applied to the estimation of the range profiles. To perform imaging in the pulse–echo mode, we transmit a so-called *probing waveform p(t)*,

$$E_T(t) = E\,p(t) \qquad 0 \le t \le T$$

The probing signal has an amplitude E and operates with a finite time duration T. The effective bandwidth of the probing signal is B.

Reflected from a single target at distance R, the echo is in the form of

$$E_R(t) = \alpha E_T(t-\tau) = \alpha E\,p(t-\tau)$$

where α denotes the attenuation due to the reflectivity of the target and propagation loss, and τ is the time delay due to the round-trip propagation from the target to the transceiver.

$$\tau = \frac{2R}{v}$$

where v is the propagation speed, and R is the distance between the target and the transceiver. Subsequently, we perform matched filtering of the returned signal with the transmitted waveform to obtain the time-delay profile,

$$\begin{aligned} R_E(t) &= E_R(t) * E_T^*(-t) \\ &= \alpha|E|^2[p(t-\tau) * p^*(-t)] \\ &= \alpha|E|^2 R_p(t-\tau) \end{aligned}$$

$R_p(t)$ is the autocorrelation of the probing signal $p(t)$. $R_p(t)$ is Hermitian symmetrical with a peak centered at $t = 0$. Thus, the result of the cross-correlation process, the range profile has a peak at $t = \tau$, with amplitude αE^2. The sharpness of the peak is largely governed by the bandwidth of the probing signal.

Because of the spectral bandwidth B and relationship between time and space $t = 2r/v$, the temporal resolution is

$$\Delta\tau = \frac{1}{B} = \frac{2\Delta R}{v}$$

This translates into the range resolution of

$$\Delta R = \frac{v}{2B}$$

For the case of multiple reflecting targets, the returned signal becomes

$$E_R(t) = \int \alpha(\tau)\, E_T(t-\tau)\, d\tau$$

And thus the time-delay profile is then in the form of

$$R_E(t) = |E|^2 \int \alpha(\tau)\, R_p(t-\tau)\, d\tau$$

4.2.1 Synthetic-Aperture Sonar Imaging

A good example of underwater acoustical imaging in the pulse–echo format is the synthetic-aperture sonar imaging, which is the acoustic equivalent of synthetic-aperture radar (SAR) imaging. The applications cover a wide range of oceanic search, surveys, and mapping. It normally functions in the reflection mode with a multi-element array. The field experiment was conducted in the San Diego Bay with a linear 10-element sonar array operating in the side-looking linear-scan model. The use of a multi-element array is to produce the redundancy for the estimation of platform motion. The platform motion consists of six parameters, of which three are associated with the

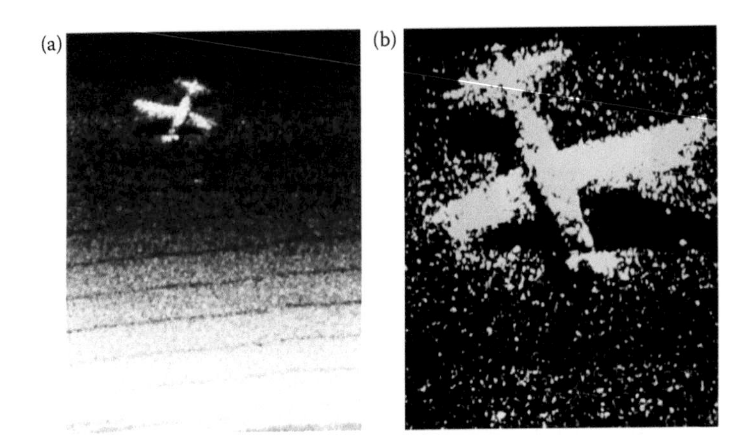

FIGURE 4.6
(a) Reconstructed image of a sunken airplane and (b) enlarged version of the image.

translational vector and another three associated with the rotation matrix in three dimensions. The synthetic aperture is one-dimensional, and thus, the final image is two-dimensional. Figure 4.6a is the reconstructed image of a sunken airplane and Figure 4.6b is the enlarged version.

4.3 Linear-Chirp Signaling

The linear-chirp imaging technique is a special case of the standard pulse–echo modality. For this case, the probing signal becomes

$$E_T(t) = E\exp\left(j\left(2\pi f_0 t + \frac{\pi B t^2}{T}\right)\right) \quad \text{for } 0 \leq t \leq T$$

where f_0 is the starting frequency and B the bandwidth of the chirp signal. The time interval $(0, T)$ is termed the signal duration and T is the so-called *chirp period*. The transmitted signal can then be rewritten as

$$E_T(t) = E\exp(j(2\pi f_0 t + \pi\beta t^2))$$

and β denotes the *slope* of the change of frequency, which is also commonly known as the chirp rate.

$$\beta = \frac{B}{T}$$

From the perspective of pulse–echo approach, the transmission waveform is equivalent to the quadratic-phase probing signal

$$p(t) = \exp(j\pi\beta t^2)$$

with amplitude E and modulated by the carrier signal $\exp(j2\pi f_0 t)$.

As it can be seen, the chirp signal is a function of varying frequency. The angular frequency of the chirp signal is a linear function of time:

$$\omega_t(t) = 2\pi f_t(t) = \frac{\partial}{\partial t}(2\pi f_0 t + \pi\beta t^2)$$
$$= 2\pi f_0 + 2\pi\beta t = 2\pi(f_0 + \beta t)$$

Alternatively, we write

$$f_t(t) = f_0 + \beta t$$

If the chirp rate is zero, $\beta = 0$, the signal degenerates back to the single-frequency coherent signal with the frequency f_0.

During the chirp period, the lower and upper bounds of the frequency are

$$f_t(0) = f_0$$
$$f_t(T) = f_0 + B$$

Thus, within one chirp period T, the frequency moves through the full bandwidth of

$$\Delta f_t = (f_0 + B) - f_0 = B$$

Consider a target reflector at a distance R. The round-trip time delay is

$$\tau = \frac{2R}{v}$$

where v is the propagation speed. The factor of two is due to the round-trip propagation. Then the echo from the target is a weighted delayed version of the transmitted signal:

$$E_R(t) = \alpha E_T(t - \tau)$$
$$= \alpha E \exp(j(2\pi f_0(t-\tau) + \pi\beta(t-\tau)^2))$$
$$= \alpha E \exp(j2\pi f_0(t-\tau))\exp(j\pi\beta(t-\tau)^2)$$
$$= \alpha E \exp(j2\pi f_0 t)\exp(-j2\pi f_0 \tau)\exp(j\pi\beta t^2)\exp(j\pi\beta\tau^2)\exp(-j2\pi\beta\tau t)$$

where α is governed by the reflectivity of the target as well as propagation loss.

The transmitted signal $E_T(t)$ is used as the reference signal to mix with the returned signal $E_R(t)$,

$$E_R(t)\, E_T^*(t) = \alpha|E|^2 \exp(-j2\pi f_0\tau)\exp(j\pi\beta\tau^2)\exp(-j2\pi\beta\tau t)$$

Due to the mixing, the five phase terms in $E_R(t)$ is reduced down to three and only one is a function of time. Now define the complex amplitude C as

$$C = \alpha|E|^2 \exp(-j2\pi f_0\tau)\exp(j\pi\beta\tau^2)$$

The output of the mixer is now in the form of a coherent function in time,

$$E_R(t)\, E_T^*(t) = C\exp(-j2\pi\beta\tau t) = C\exp(-j2\pi(\beta\tau)t)$$

with complex amplitude C and one single frequency

$$\hat{f} = \beta\tau = \frac{B(2R/v)}{T}$$

$$= \left(\frac{2B}{vT}\right)R = \left(\frac{2}{v}\beta\right)R$$

This means, if we Fourier transform the output of the mixer, the spectrum shows one peak at $\hat{f}$ and the amplitude of the peak is C. In addition, the peak frequency is linearly related to the range distance. Therefore, the Fourier transform of the output of the mixer gives the range profile of the target region and the range distance of a target reflector is related to the frequency index,

$$R = \left(\frac{vT}{2B}\right)\hat{f}$$

Thus, the Fourier spectrum of the output of the mixer can be regarded as a scaled version of the range profile, which is the scaling factor ($vT/2B$). For a time signal with bandwidth B, the temporal resolution is

$$\Delta\tau = \frac{1}{B} = \frac{2\Delta R}{v}$$

This translates into the range resolution of

$$\Delta R = \frac{v}{2B}$$

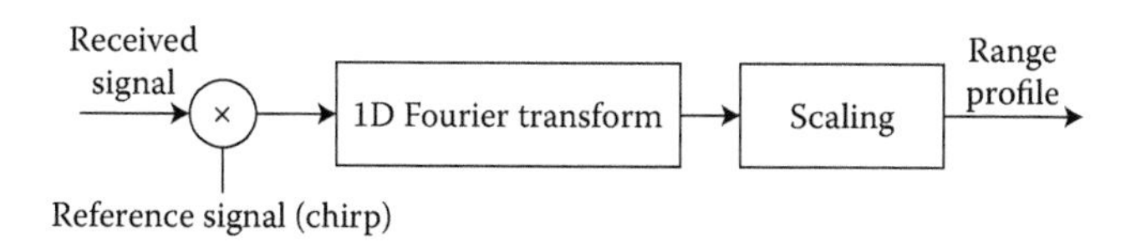

FIGURE 4.7
Range estimation with linear-chirp waveforms.

which is governed by the propagation speed v and bandwidth B. This leads to exactly the same result from the pulse–echo model. Figure 4.7 shows the diagram for range estimation with linear-chirp waveforms.

4.4 Step-Frequency FMCW Ultrasound Imaging

The step-frequency FMCW imaging technique is the hybrid version of the pulse–echo and chirp signal modalities. It utilizes a collection of CW pulses as the probing signals. At each frequency cycle, it is operating with the same concept and in the same format as the pulse–echo system. With a full collection of frequency steps, it creates the bandwidth similar to the function of use of the chirp waveforms.

During a signaling cycle, the transmitter sends out a complete sequence of N coherent signals, stepping through a defined frequency band B with frequency increment Δf.

$$f = f_0 + k\Delta f \quad \text{where } k = 0, 1, 2, \dots, N-1$$

During each frequency step, the system functions exactly like a CW system. So, from this perspective, the probing waveforms of the step-frequency FMCW modality can be regarded as an organized sequence of CW signals.

For simplicity, the transmitted signal can be written in the phasor form of

$$E_T(t) = E\exp(j2\pi ft)$$

For each frequency step, this is equivalent to operating in the pulse–echo mode with a finite-duration pulse as the probing function and a carrier frequency f. In other words, it is to repeat the pulse–echo mode N times, with N different carrier frequencies. The pulse period is sufficiently long to be considered as a CW format. That is why FMCW imaging systems typically require longer data-acquisition time. In return, the quality of the final images is better due to lower noise level.

For a target at a range distance R, the delay due to round-trip travel is

$$\tau = \frac{2R}{v}$$

where v is the propagation speed. Then, responding to this target, the reflected signal detected by the receiver is in the form of a weighted and delayed version of the transmitted waveform,

$$E_R(t) = \alpha E \exp(j2\pi f(t-\tau)) = \alpha E \exp\left(j2\pi f\left(t - \frac{2R}{v}\right)\right)$$

where α is the weighting for the target reflectivity and propagation loss, and the delay τ is due to the round-trip travel. At each frequency step, through the quadrature detector, the received signal becomes,

$$\begin{aligned} E_R(t)\, E_T^*(t) &= \alpha |E|^2 \exp(-j2\pi f \tau) \\ &= \alpha |E|^2 \exp\left(-j2\pi f\left(\frac{2R}{v}\right)\right) \end{aligned}$$

Note that, after the demodulation, it is no longer a function of time. For each frequency, the output after the quadrature receiver is a complex scalar. Through the N frequency steps, a complete illumination cycle produces an N-point sequence $\{E(k)\}$ from the demodulated received waveforms:

$$\begin{aligned} E(k) &= E_R(t)\, E_T^*(t) \\ &= \alpha |E|^2 \exp\left(-j2\pi(f_0 + k\,\Delta f)\left(\frac{2R}{v}\right)\right) \\ &= \alpha |E|^2 \exp\left(-j2\pi f_0\left(\frac{2R}{v}\right)\right)\exp\left(-j2\pi k \Delta f\left(\frac{2R}{v}\right)\right) \end{aligned}$$

The term $\exp(-j2\pi k\Delta f(2R/v))$ is the only function of the frequency-stepping index k. To analyze this further, we match the core part of the sequence against the kernel of the fast Fourier transform (FFT) operator.

$$\exp\left(-j2\pi k \Delta f\left(\frac{2R}{v}\right)\right) = \exp\left(\frac{-j2\pi nk}{N}\right)$$

The matching results in a simple relationship,

$$\frac{n}{N} = 2\Delta f \frac{R}{v}$$

It is then simplified down to a linear relationship between the FFT index n and the target range R,

$$n = N\left(2\Delta f \frac{R}{v}\right) = 2(N\Delta f)\frac{R}{v} = \frac{2B}{v}R$$

where B is the bandwidth of the waveform, defined as $B = N\Delta f$. This implies that the term $E(k)$ can be written in the form of

$$E(k) = E_R(t)\, E_T^*(t) = \alpha\,|E|^2 \exp\left(-j2\pi f_0\left(\frac{2R_0}{v}\right)\right)\exp\left(\frac{-j2\pi n_0 k}{N}\right)$$

which is corresponding to a point sequence in the frequency domain

$$\begin{aligned}\alpha(n) &= \alpha\delta(n - n_0)\\ &= \alpha\delta\left(n - \frac{2B}{v}R_0\right)\end{aligned}$$

with the linear correspondence of

$$n_0 = \frac{2B}{v}R_0$$

For a general range profile, the range distribution can be generalized to the form of $\alpha(n)$. Then the range profile $\alpha(n)$ and the sequence $E(k)$ are directly related as a Fourier-transform pair that each index n of the spectral sequence is corresponding to a range distance R,

$$\text{range distance} = R = \frac{v}{2B}n$$

Since the index n is an integer, the scaling factor $v/2B$ represents the increment of the range profile, which is often referred to as the range resolution,

$$\text{range resolution} = \Delta R = \frac{v}{2B}$$

The resolution in the range direction is governed by the operating bandwidth of the FMCW system. Wider bandwidth produces improved resolving capability in a linear manner. Thus, for the FMCW system, the range estimation process can be simplified down to an inverse FFT operation, followed by a linear scaling operation, which is a more effective approach than the conventional matched filtering method.

The complex term $\alpha(n)|E|^2\exp(-j2\pi f_0(2R/v))$ can be reorganized into the form of

$$\alpha(n)|E|^2\exp\left(-j2\pi f_0\left(\frac{2R}{v}\right)\right) = \alpha(n)|E|^2\exp\left(-j2\pi\left(\frac{f_0}{B}\right)n\right)$$

Thus, the range estimation process can be summarized as

1. Demodulate the received waveforms to obtain the N-point data sequence $E(k)$.
2. FFT the data sequence to obtain preliminary range profile.
3. Normalize the preliminary range profile by removing the complex amplitude term $|E|^2\exp(-j2\pi(f_0/B)n)$.
4. Rescale the resultant sequence with the resolution unit at $v/2B$.

Figure 4.8 shows the range estimation process with step-frequency FMCW waveforms.

Note that the multiplication factor $\exp(-j2\pi(f_0/B)n)$ can be achieved by padding k_0 to the front of the sequence $E(k)$, if we can identify a k_0 such that

$$\frac{k_0}{N} = \frac{f_0}{B}$$

which leads to

$$k_0 = \frac{f_0}{\Delta f}$$

As indicated, the resolution in the range direction is governed by the bandwidth of the probing signal. The resolution analysis produces exactly the same result for pulse–echo, chirp signals, and step-frequency FMCW waveforms. It is equivalent to the concept of time-bandwidth product. Similar result can be obtained in the form of the Cramer–Rao bound, which is formulated in terms of the RMS bandwidth instead.

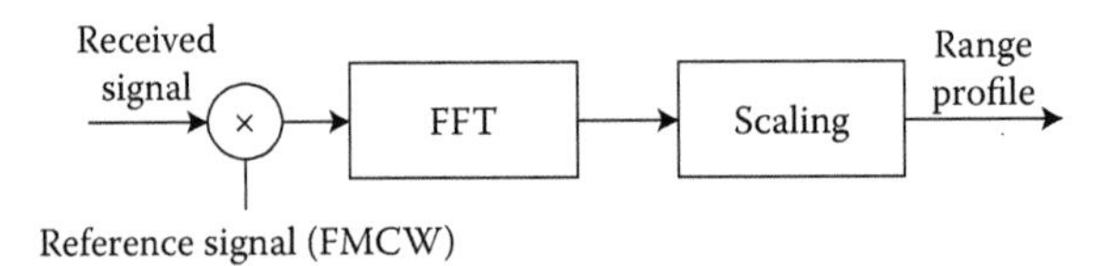

FIGURE 4.8
Range estimation with step-frequency FMCW waveforms.

4.4.1 Superposition of the Range Profiles

At each receiver position (x_m, y_m, z_m), we compute the range profile $\hat{s}(x_m, y_m, z_m; r)$. For simplicity, we denote it as $\hat{s}_m(r)$. This represents the range profile estimated at the mth receiver position (x_m, y_m, z_m),

$$\hat{s}_m(r) = \hat{s}(x_m, y_m, z_m; r)$$

Conceptually, we can spread a range profile $\hat{s}_m(r)$ over the region of interest to form a sub-image $\hat{s}_m(x, y, z)$, where (x, y, z) is the pixel position within the region of interest. This process can also include an additional weighting $w_m(r)$. With the weighting, the range profile can be modified in the form of $w_m(r)\ \hat{s}_m(r)$ prior to the formation of the sub-image $\hat{s}_m(x, y, z)$.

It is also common to expand the weighting function $w(r)$ to incorporate

1. The range factor in Green's function
2. Normalization factor for limited aperture size
3. Beam patterns
4. Range compensation for the propagation loss due to distance

The formation of the mth sub-image $\hat{s}_m(x, y, z)$ is a simple conversion from the mth range profile $\hat{s}_m(r)$. Suppose the mth profile is corresponding to the transmitter located at (x_t, y_t, z_t) and receiver at (x_r, y_r, z_r). For each position (x, y, z) in the region of interest, we first compute the distances to the transmitter and to the receiver, respectively,

$$\begin{aligned} r &= r_1 + r_2 \\ &= [(x - x_t)^2 + (y - y_t)^2 + (z - z_t)^2]^{1/2} \\ &\quad + [(x - x_r)^2 + (y - y_r)^2 + (z - z_r)^2]^{1/2} \end{aligned}$$

Then we use the sum of the travel distances, $r = r_1 + r_2$, to seek and assign the complex amplitude from the mth range profile $\hat{s}_m(r)$ at $r = r_1 + r_2$ to the pixel of $\hat{s}_m(x, y, z)$ to form the mth sub-image

$$\hat{s}_m(x, y, z) = \hat{s}_m(r_1 + r_2)$$

Then subsequent step of the image reconstruction procedure is the superposition of the M sets of range profiles to form the final image

$$\hat{s}(x,\ y,\ z) = \frac{1}{M} \sum_{m=1}^{M} \hat{s}_m(x, y, z)$$

This is equivalent to summing up the values extracted from all the range profiles corresponding to the location (x, y, z).

This superposition procedure does not require extensive computation. For each pixel position (x, y, z), we compute r_1 and r_2 corresponding to each range profile, given its transmitter and receiver positions. From each range profile, the value of range profile at range bin of $r = r_1 + r_2$ is selected. The value of the final image at (x, y, z), is the average of the M selected values from M range profiles.

To improve the computation efficiency, the values of r_1 and r_2 are precomputed and stored for each pixel to form a look-up table. For each pixel position, the look-up table contains M range-bin addresses corresponding to the M range profiles. Thus, main task of this superposition procedure becomes a simple look-up and retrieve process.

4.4.2 FMCW Medical Ultrasound Imaging

One of the direct applications of the FMCW imaging technique is in the area of medical ultrasound. Because this method conducts the estimation of the range profiles first, the geometrical configuration of the transceiver arrays becomes flexible and reconfiguration is feasible. Figure 4.9 is the microflexible ultrasound transceiver array prototype for laboratory experiments. Figure 4.10 is the data-acquisition hardware of the imaging device.

The length of the FMCW sequence directly defines the bandwidth of the range profile, which governs the resolution in the range direction. Thus, at a given frequency increment, longer FMCW data sequences translate into improved resolution. Figure 4.11a–d shows the reconstructed images of five point targets from FMCW data of different lengths, illustrating the improvement of resolution as the bandwidth increases. The full circular aperture is employed to provide illustration with uniform resolution in all directions.

With simple modifications, the system can perform imaging outward, depending on the physical constraints and applications. Figure 4.12 shows an outward-looking array for biomedical imaging of the prostate region. Figure 4.13 is the reconstructed image of nine small point targets.

FIGURE 4.9
Microflexible ultrasound transceiver array prototype.

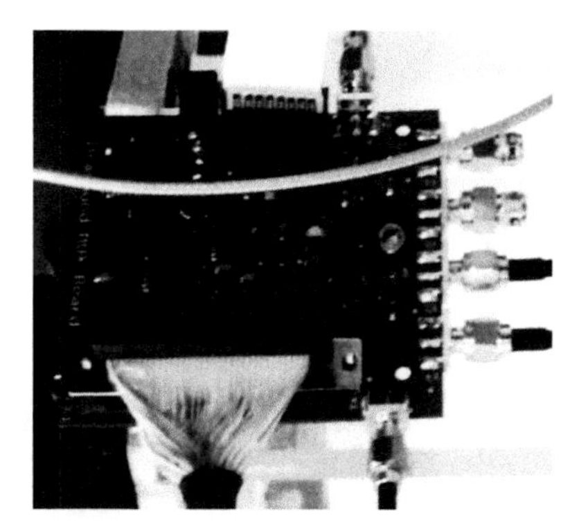

FIGURE 4.10
Data-acquisition hardware of the ultrasound imaging device.

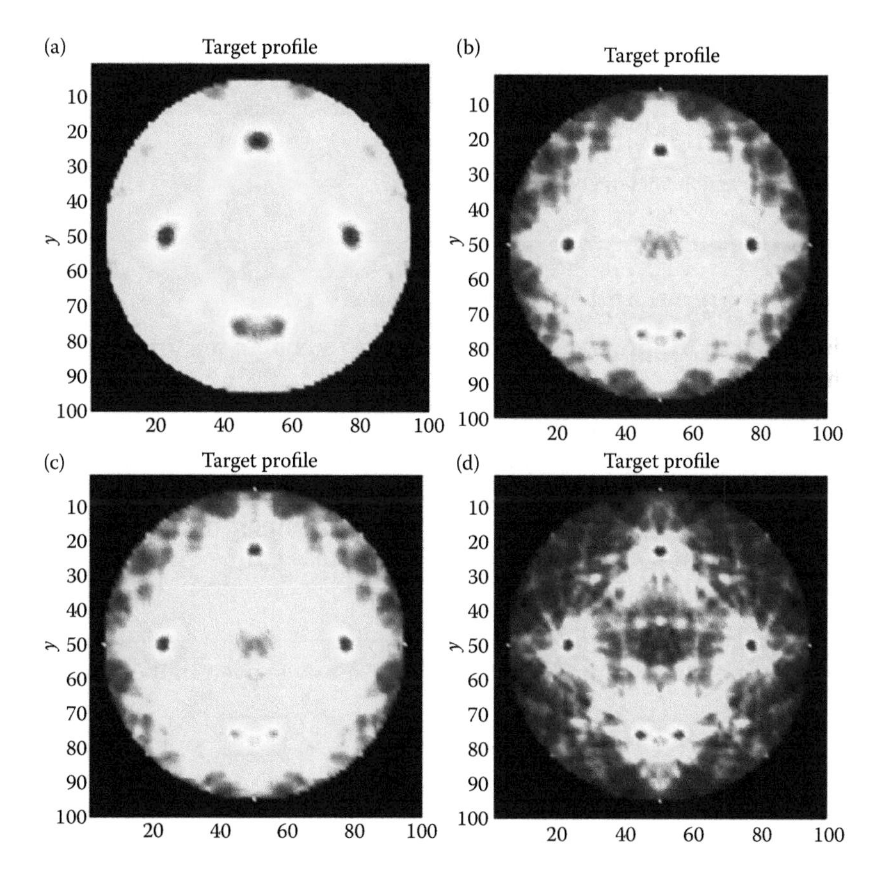

FIGURE 4.11
(a–d) Images of five point targets corresponding to various FMCW bandwidth.

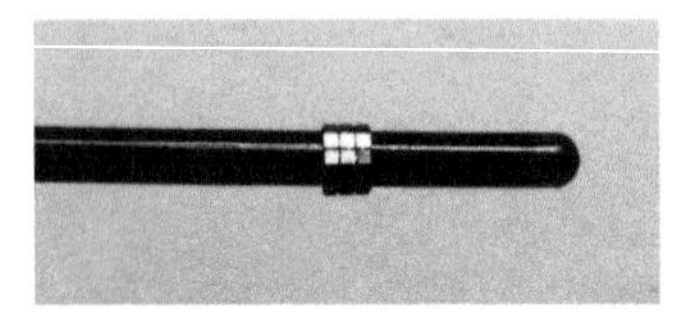

FIGURE 4.12
Outward-looking array.

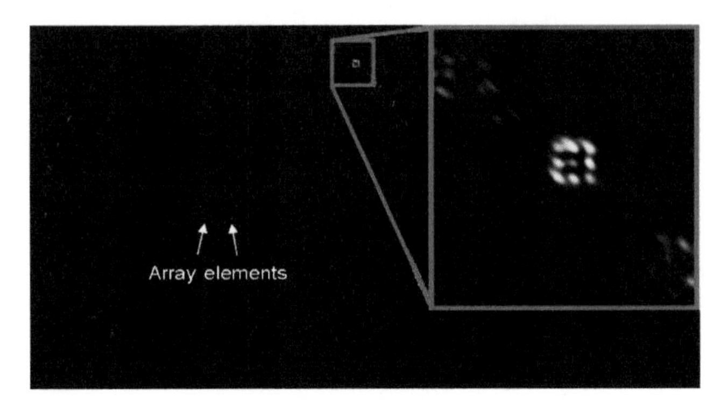

FIGURE 4.13
Image of nine small point targets by the array.

4.5 Chirp Signal and Fresnel Phase Pattern

As indicated in the previous section, the chirp signal in the time domain is in the form of

$$E_T(t) = E\exp(j(2\pi f_0 t + \pi\beta t^2))$$

The term $\exp(j2\pi f_0 t)$ is the carrier component associated with the physical propagation of the probing waveform. The core component of the chirp signal is the core component with the quadratic-phase term

$$E(t) = \exp(j\pi\beta t^2)$$

From a different section, the Fresnel phase mask is given in the form of

$$p(x, y) = \exp\left(\frac{j\pi(x^2 + y^2)}{\lambda z_0}\right)$$

It is circularly symmetrical, and can be rewritten in the polar form as

$$p(r) = \exp\left(\frac{j\pi r^2}{\lambda z_0}\right)$$

where $r = (x^2 + y^2)^{1/2}$. So, the Fresnel phase pattern can be regarded as a two-dimensional version of the chirp signal in the space domain.

If we give the Fresnel phase pattern a spatial offset of (x_0, y_0), it becomes

$$p(x - x_0, y - y_0) = \exp\left(\frac{j\pi((x - x_0)^2 + (y - y_0)^2)}{\lambda z_0}\right)$$

Now, we mix it with the original phase pattern as the reference waveform, it is then simplified to

$$\begin{aligned}
p(x-x_0, y-y_0)\, p^*(x,y) &= \exp\left(\frac{j\pi((x-x_0)^2+(y-y_0)^2)}{\lambda z_0}\right)\exp\left(\frac{-j\pi(x^2+y^2)}{\lambda z_0}\right) \\
&= \exp\left(\frac{j\pi((x^2+x_0^2-2x_0x)+(y^2+y_0^2-2y_0y))}{\lambda z_0}\right) \\
&\quad \times \exp\left(\frac{-j\pi(x^2+y^2)}{\lambda z_0}\right) \\
&= \exp\left(\frac{j\pi(x_0^2+y_0^2)}{\lambda z_0}\right)\exp\left(\frac{-j2\pi(x_0x+y_0y)}{\lambda z_0}\right)
\end{aligned}$$

The complex phase term $\exp(j\pi(x_0^2 + y_0^2)/\lambda z_0)$ is not a function of the space variables x and y. Thus it can be considered as part of the complex amplitude. And the second term, $\exp(-j2\pi(x_0x + y_0y)/\lambda z_0)$, a single-frequency spatial distribution with the frequency vector

$$\begin{aligned}
[f_x, f_y] &= \left[\frac{-x_0}{\lambda z_0}, \frac{-y_0}{\lambda z_0}\right] \\
&= \left(\frac{-1}{\lambda z_0}\right)[\, x_0, y_0]
\end{aligned}$$

This shows that the spatial frequency of the single-frequency pattern is linearly related to the spatial offset. Hence, if we perform Fourier transform on the result of the mixer, the spatial-frequency spectrum contains one single peak

$$\begin{aligned}
\boldsymbol{F}\{p(x - x_0, y - y_0)\, p^*(x,y)\} &= \boldsymbol{F}\left\{\exp\left(\frac{j\pi(x_0^2 + y_0^2)}{\lambda z_0}\right)\exp\left(\frac{-j2\pi(x_0x + y_0y)}{\lambda z_0}\right)\right\} \\
&= \exp\left(\frac{j\pi(x_0^2 + y_0^2)}{\lambda z_0}\right)\delta\left(\frac{f_x + x_0}{\lambda z_0}, \frac{f_y + y_0}{\lambda z_0}\right)
\end{aligned}$$

The spectral peak is located at $(-x_0/\lambda z_0, -y_0/\lambda z_0)$ with the amplitude $\exp(j\pi(x_0^2+y_0^2)/\lambda z_0)$. This simple technique can be applied to effective detection and estimation of spatial offset and misalignment. Figure 4.14 shows the displacement estimation with Fresnel phase patterns.

4.5.1 Implementation by FFT

Because of the use of Fresnel and Fraunhofer approximations, chirp waveforms, and step-frequency FMCW probing signals, the matched filtering procedure is simplified to the level of a Fourier-transform operation, which is implemented with the use of the discrete Fourier transform/fast Fourier transform (DFT/FFT) routines. In this section, the physical meanings of the parameters of the DFT implementation are examined.

Assume the size of the source is D_1, and the size of the receiving aperture is D_2. The number of wavefield data samples over the aperture is N. This suggests that the sample spacing of the wavefield is $\Delta x = D_2/N$. After the reconstruction, the sample spacing of the image over the source region is $\Delta x' = D_1/N$.

If we approximate $x = k\Delta x = kD_2/N$ and $x' = n\Delta x' = nD_1/N$, the kernel associated with the Fresnel phase pattern becomes

$$\exp\left(\frac{-j2\pi x'x}{\lambda z_0}\right) = \exp\left(\frac{-j2\pi(n\,D_1/N)(k\,D_2/N)}{\lambda z_0}\right)$$

$$= \exp\left(-j2\pi\left(\frac{nk}{N}\right)\left(\frac{D_1D_2}{N\lambda z_0}\right)\right)$$

Now we match it with the DFT kernel:

$$\exp\left(-j2\pi\left(\frac{nk}{N}\right)\left(\frac{D_1D_2}{N\lambda z_0}\right)\right) = \exp\left(\frac{-j2\pi nk}{N}\right)$$

This leads to the simple relationship:

$$\frac{D_1D_2}{N\lambda z_0} = 1$$

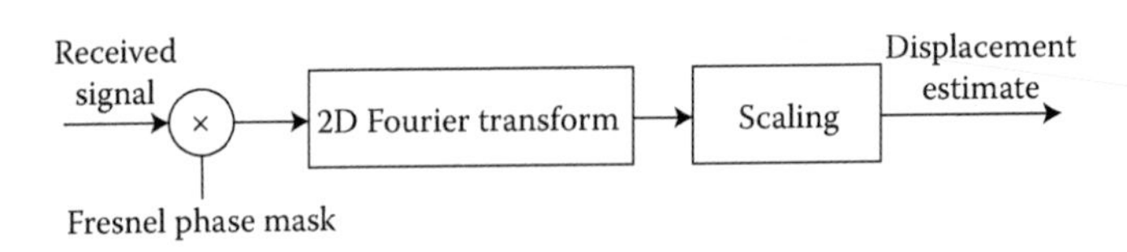

FIGURE 4.14
Displacement estimation with Fresnel phase patterns.

This relationship can be observed from two interesting perspectives. The first is

$$\Delta x = \frac{D_2}{N} = \frac{\lambda z_0}{D_1}$$
$$= \left(\frac{\lambda}{2}\right)\left(\frac{2z_0}{D_1}\right) = \frac{(\lambda/2)}{(D_1/2z_0)}$$
$$= \left(\frac{\lambda}{2}\right)\frac{1}{\tan(\phi/2)} \approx \left(\frac{\lambda}{2}\right)\frac{1}{\sin(\varphi/2)}$$

where ϕ denotes the angular span of the source region from the perspective of the aperture. This equation suggests small sample spacing for larger source size, and the smallest wavefield sample spacing is half wavelength.

The second one is the dual version. The equation can be rearranged into the form of

$$\Delta x' = \frac{D_1}{N} = \frac{\lambda z_0}{D_2}$$
$$= \left(\frac{\lambda}{2}\right)\frac{1}{\tan(\theta/2)} \approx \left(\frac{\lambda}{2}\right)\frac{1}{\sin(\theta/2)}$$

where θ denotes the angular span of the aperture from the perspective of the source. It implies higher resolution for larger aperture size, and the limit of the resolving capability is half wavelength. This relationship matches the formula of Rayleigh resolution limit, exactly.

Another interesting application is the formation of the range profile from the FMCW data sequence. Assume the bandwidth of the FMCW coverage is B, and the maximum range is D. Suppose the number of FMCW frequency steps is N. This suggests that the frequency step size is $\Delta f = B/N$.

After the reconstruction, the sample spacing of the range profile is $\Delta r = D/N$.

$$\exp(-j2\pi f\tau) = \exp(-j2\pi(k\Delta f)(n\Delta\tau))$$
$$= \exp\left(-j2\pi\left(\frac{kB}{N}\right)\left(\frac{n\Delta r}{v}\right)\right)$$
$$= \exp\left(-j2\pi\left(\frac{nk}{N}\right)\left(\frac{N\Delta f\Delta r}{v}\right)\right)$$

Similarly, this matches with the DFT kernel

$$\exp\left(-j2\pi\left(\frac{nk}{N}\right)\left(\frac{N\Delta f\Delta r}{v}\right)\right) = \exp\left(\frac{-j2\pi nk}{N}\right)$$

This leads to the simple relationship

$$\frac{N\Delta f \Delta r}{v} = 1$$

After rearrangement, it gives the range resolution as

$$\Delta r = \frac{v}{N\Delta f} = \frac{v}{B}$$

This matched the results from previous sections. If we consider active ranging modality, it gives an extra factor of two for round-trip propagation.

The same equation also defines the size of the frequency steps as

$$\Delta f = \frac{v}{N\Delta r} = \frac{v}{D}$$

So, the ratio of propagation speed v and maximum range distance D governs the frequency step size of the FMCW waveform sequence. An extra factor of two also appears when the system operates in the active mode.

The general formulation of multi-frequency backward propagation algorithm for image reconstruction is a superposition of a collection of coherent sub-images. In terms of the computational structure, the backward propagation algorithm consists of two basic steps. The first is the coherent backward propagation, and then followed by a complex superposition procedure. The coherent backward propagation is a matched filtering process, and thus structured array configuration is crucial to the implementation of the matched filtering in the form of a convolution integral.

If we reverse the order of the multi-frequency backward propagation procedure, we can implement the procedure in the range direction first by estimating the range profiles at various receiver positions. Each range profile can form a sub-image, and the final image is the superposition of the sub-images. This format allows us to relax the constraints of array configuration, such that the concept of dynamically reconfigurable arrays becomes realizable.

The estimation of the range profile is also a matched filtering process. When chirp or step-frequency FMCW waveforms are employed, the matched filtering process can be simplified down to a Fourier-transform operation. Nonetheless, the resolution in the range direction remains the same for pulse–echo, linear chirp, and step-frequency FMCW illumination.

5

Resolution Enhancement and Motion Estimation

In the field of imaging technology, the system analysis tasks do not conclude at the stage of image reconstruction. There are numerous important tasks subsequent to the image formation process. In this chapter, several typical exercises are documented to illustrate the processes of performance evaluation and enhancement subsequent to image formation.

This chapter consists of four sections, grouped into two parts. The first part is devoted to enhancement techniques. The first technique is for the reduction or removal of phase errors introduced to the data-acquisition process for the use of quadrature receivers. This degradation factor is common in coherent or step-frequency FMCW (frequency-modulated continuous wave) imaging systems. The error is often overlooked because it cannot be modeled as additive noise. The second technique is the formation of an enhancement operator, with wavefield statistics, based on the fundamental concept of backward-propagation image formation.

The second half of the chapter is allocated to motion estimation techniques. Displacement estimation is an important task for many imaging applications. To be thorough and complete, this chapter covers both the parameter-based and image-based techniques in both space and frequency domain.

5.1 Quadrature-Receiver Phase Errors and Correction

5.1.1 Data Acquisition of the Step-Frequency FMCW Systems

The complex FMCW (frequency-modulated continuous wave) wavefield data sequence from the quadrature receivers is in the form of

$$x(k) = A_k \exp(j\theta_k)$$

corresponding to the N coherent illumination-frequency steps,

$$\omega = \omega_k \quad k = 0,1,2,\ldots,N-1$$

Normally, the range profile $R(n)$ can then be obtained as the fast Fourier transform (FFT) of the data sequence $\{x(k)\}$.

$$R(n) = FFT\{x(k)\}$$

The complex wavefield data $A\exp(j\theta)$ is the phasor version of the returned signal $A\cos(\omega t + \theta)$. For an FMCW system, the data acquisition is performed by a collection of N quadrature receivers, each with a pair of orthogonal reference signals,

$$r_I(t) = \cos(\omega t)$$

and

$$r_Q(t) = \sin(\omega t)$$

This pair of coherent reference signals maintains a 90° phase offset. After mixing with the reference signals, the two channels of the returned signal become

$$A\cos(\omega t + \theta)\cos(\omega t) = \frac{A}{2}\cos(2\omega t + \theta) + \frac{A}{2}\cos(\theta)$$

and

$$A\cos(\omega t + \theta)\sin(\omega t) = \frac{A}{2}\sin(2\omega t + \theta) - \frac{A}{2}\sin(\theta)$$

After lowpassing, the outputs from the I/Q channels of the quadrature receiver are in the form of

$$g_I = \frac{A}{2}\cos(\theta)$$

and

$$g_Q = -\frac{A}{2}\sin(\theta)$$

Subsequent to amplitude adjustments, the real and imaginary components of the signal become

$$\mathrm{Re} = A\cos(\theta)$$

and

$$Im = A\sin(\theta)$$

This leads to the complex wavefield data sequence

$$A_k \exp(j\theta_k) = A_k \cos(\theta_k) + jA_k \sin(\theta_k)$$

5.1.2 Quadrature-Receiver Phase Errors

In practice, the hardware electronics of the quadrature receiver is not able to maintain the 90° phase offset precisely and, as a result, often introduces a phase error $\Delta\theta$,

$$r_I(t) = \cos\left(\omega t + \frac{\Delta\theta}{2}\right)$$

and

$$r_Q(t) = \sin\left(\omega t - \frac{\Delta\theta}{2}\right)$$

For symmetry, we distribute the phase error evenly to the reference signal pair. As expected, this phase error propagates into the output channels,

$$A\cos(\omega t + \theta)\cos\left(\omega t + \frac{\Delta\theta}{2}\right) = \frac{A}{2}\cos\left(2\omega t + \theta + \frac{\Delta\theta}{2}\right) + \frac{A}{2}\cos\left(\theta - \frac{\Delta\theta}{2}\right)$$

$$A\cos(\omega t + \theta)\sin\left(\omega t - \frac{\Delta\theta}{2}\right) = \frac{A}{2}\sin\left(2\omega t + \theta - \frac{\Delta\theta}{2}\right) - \frac{A}{2}\sin\left(\theta + \frac{\Delta\theta}{2}\right)$$

After lowpassing, the outputs from the *I/Q* channels now become

$$g_I = \frac{A}{2}\cos\left(\theta - \frac{\Delta\theta}{2}\right)$$

and

$$g_Q = -\frac{A}{2}\sin\left(\theta + \frac{\Delta\theta}{2}\right)$$

Due to the phase error, the complex wavefield data is now in the form of

$$A\cos\left(\theta-\frac{\Delta\theta}{2}\right)+jA\sin\left(\theta+\frac{\Delta\theta}{2}\right)$$

$$=\frac{A}{2}\left[\exp\left(j\left(\theta-\frac{\Delta\theta}{2}\right)\right)+\exp\left(-j\left(\theta-\frac{\Delta\theta}{2}\right)\right)\right]$$

$$+\frac{A}{2}\left[\exp\left(j\left(\theta+\frac{\Delta\theta}{2}\right)\right)-\exp\left(-j\left(\theta+\frac{\Delta\theta}{2}\right)\right)\right]$$

$$=\frac{A}{2}\left[\exp\left(j\frac{\Delta\theta}{2}\right)+\exp\left(-j\frac{\Delta\theta}{2}\right)\right]\exp(j\theta)$$

$$+\frac{A}{2}\left[\exp\left(j\frac{\Delta\theta}{2}\right)-\exp\left(-j\frac{\Delta\theta}{2}\right)\right]\exp(-j\theta)$$

$$=A\cos\left(\frac{\Delta\theta}{2}\right)\exp(j\theta)+A\sin\left(\frac{\Delta\theta}{2}\right)\exp(-j\theta)$$

Upon close examination, instead of the original form of $A_k \exp(j\theta_k)$, the data sequence becomes

$$\hat{x}(k)=\left\{A_k\cos\left(\frac{\Delta\theta_k}{2}\right)\exp(j\theta_k)+A_k\sin\left(\frac{\Delta\theta_k}{2}\right)\exp(-j\theta_k)\right\}$$

$$=\left\{\cos\left(\frac{\Delta\theta_k}{2}\right)A_k\exp(j\theta_k)\right\}+\left\{\sin\left(\frac{\Delta\theta_k}{2}\right)A_k\exp(-j\theta_k)\right\}$$

$$=\cos\left(\frac{\Delta\theta_k}{2}\right)x(k)+\sin\left(\frac{\Delta\theta_k}{2}\right)x^*(k)$$

It can be seen that, due to the phase error, the FMCW sequence now has two components. One is $\cos(\Delta\theta_k/2)x(k)$, which is the original sequence $x(k)$ modulated by $\cos(\Delta\theta_k/2)$. Since $x(k)$ is corresponding to the range profile $R(n)$, the term $\cos(\Delta\theta_k/2)x(k)$ implies that the range profile is degraded by the modulation term $\cos(\Delta\theta_k/2)$. The other term is $\sin(\Delta\theta_k/2)x^*(k)$, which is the conjugate of $x(k)$ modulated by $\sin(\Delta\theta_k/2)$.

5.1.3 Estimation of the Phase Errors

Although degraded by the modulation term $\cos(\Delta\theta_k/2)$, the first component $\cos(\Delta\theta_k/2)x(k)$ produces a range profile in the correct interval in space. On the other hand, $x^*(k)$ gives the flipped version of the range profile, due to the conjugation. Thus, the second term $\sin(\Delta\theta_k/2)x^*(k)$ produces an artifact range distribution in the negative-range direction. This artifact has lower magnitude because of the modulation term $\sin(\Delta\theta_k/2)$, which is relatively small when the phase errors are not large.

Now, we examine the real and imaginary components of the data sequence. As it was mentioned, the term

$$x(k) = A_k \exp(j\theta_k)$$

is responsible for the range profile in the positive-range direction. However, the conjugate term

$$x^*(k) = A_k \exp(-j\theta_k)$$

produces a distribution in the negative-range direction. In an imaging system, all target range distances are positive. Thus, it can be seen that the term $\sin(\Delta\theta_k/2)A_k \exp(-j\theta_k)$ is not valid and exists solely because of the phase errors.

To illustrate the phase-error estimation procedure, we here rewrite the real and imaginary components of the data sequence in the form of

$$A\cos\left(\theta - \frac{\Delta\theta}{2}\right) = \frac{A}{2}\exp\left(-j\frac{\Delta\theta}{2}\right)\exp(j\theta) + \frac{A}{2}\exp\left(j\frac{\Delta\theta}{2}\right)\exp(-j\theta)$$

$$jA\sin\left(\theta + \frac{\Delta\theta}{2}\right) = \frac{A}{2}\exp\left(j\frac{\Delta\theta}{2}\right)\exp(j\theta) - \frac{A}{2}\exp\left(-j\frac{\Delta\theta}{2}\right)\exp(-j\theta)$$

Thus, corresponding to the positive-range direction, the single-sideband components of the real and imaginary parts are

$$\hat{g}_{ISS}(k) = \frac{A}{2}\exp\left(\frac{-j\Delta\theta_k}{2}\right)\exp(j\theta_k)$$

and

$$\hat{g}_{QSS}(k) = \frac{A}{2}\exp\left(\frac{j\Delta\theta_k}{2}\right)\exp(j\theta_k)$$

Thus, the phase error can be estimated from the ratio of the single-sideband components.

$$\frac{\hat{g}_{QSS}(k)}{\hat{g}_{ISS}(k)} = \exp(j\Delta\theta_k)$$

5.1.4 Correction Procedure

Once the phase errors are estimated, wavefield data correction can be performed for resolution improvement of the range profile. For each frequency, the relationship between the quadrature data, with and without the phase error, can be formulated as

$$A\cos\left(\theta-\frac{\Delta\theta}{2}\right)=\cos\left(\frac{\Delta\theta}{2}\right)A\cos(\theta)+\sin\left(\frac{\Delta\theta}{2}\right)A\sin(\theta)$$

and

$$A\sin\left(\theta+\frac{\Delta\theta}{2}\right)=\sin\left(\frac{\Delta\theta}{2}\right)A\cos(\theta)+\cos\left(\frac{\Delta\theta}{2}\right)A\sin(\theta)$$

More effectively, the relationship can be written in the matrix form

$$\begin{bmatrix} A\cos\left(\theta-\frac{\Delta\theta}{2}\right) \\ A\sin\left(\theta+\frac{\Delta\theta}{2}\right) \end{bmatrix}=\begin{bmatrix} \cos\left(\frac{\Delta\theta}{2}\right) & \sin\left(\frac{\Delta\theta}{2}\right) \\ \sin\left(\frac{\Delta\theta}{2}\right) & \cos\left(\frac{\Delta\theta}{2}\right) \end{bmatrix}\begin{bmatrix} A\cos(\theta) \\ A\sin(\theta) \end{bmatrix}$$

The correction procedure can then be achieved by a simple 2×2 matrix multiplication

$$\begin{bmatrix} A\cos(\theta) \\ A\sin(\theta) \end{bmatrix}=\frac{1}{\cos(\Delta\theta)}\begin{bmatrix} \cos\left(\frac{\Delta\theta}{2}\right) & -\sin\left(\frac{\Delta\theta}{2}\right) \\ -\sin\left(\frac{\Delta\theta}{2}\right) & \cos\left(\frac{\Delta\theta}{2}\right) \end{bmatrix}\begin{bmatrix} A\cos\left(\theta-\frac{\Delta\theta}{2}\right) \\ A\sin\left(\theta+\frac{\Delta\theta}{2}\right) \end{bmatrix}$$

For N frequency steps, this correction process is conducted N times. Because of additive noise in the data-acquisition process, the ratio of the signal-sideband data sequences may not be phase-only. For this case, the estimation–correction procedure can be performed iteratively. The key indicator of the existence of the quadrature-receiver phase errors is the range profile in the negative-range direction. To achieve removal of the phase errors, the process can be conducted recursively until the component in the negative-range direction is eliminated or minimized. Figure 5.1 shows the range profile of two point targets, with quadrature phase errors. A clear indicator is the false target profile in the negative-frequency sideband. The mean of the phase errors is 15° with standard deviation of 5°. Figure 5.2 shows the range profile after the phase correction procedure.

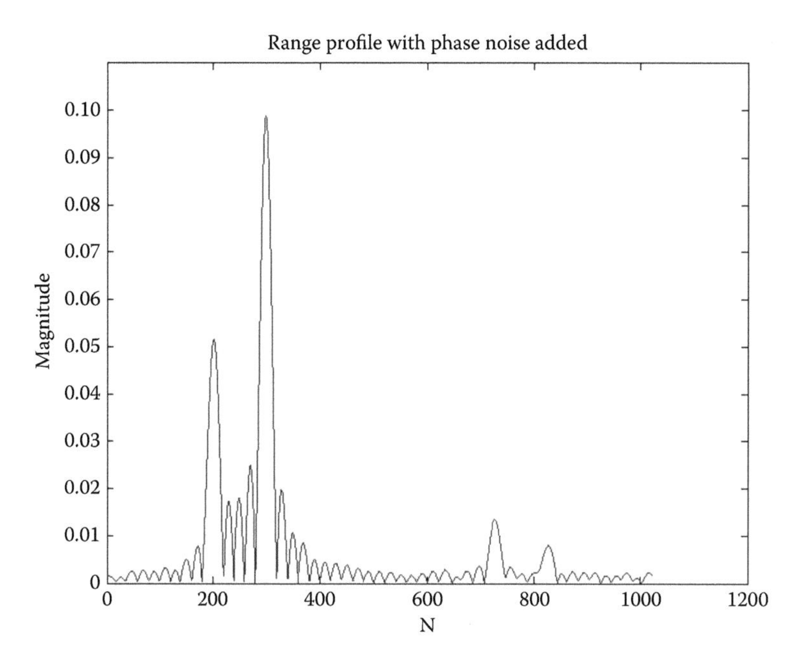

FIGURE 5.1
Range profile of two point targets, with quadrature phase errors (error mean = 15°, standard deviation = 5°).

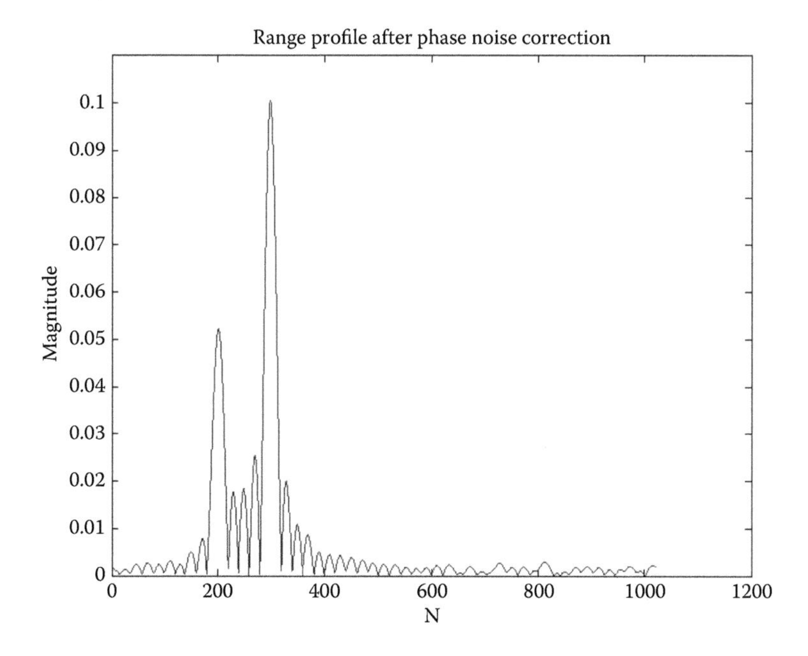

FIGURE 5.2
Range profile after phase correction.

Because of the additive noise in the detected wavefield samples, the estimate of the phase errors may not be exact. Thus, the error estimation–correction procedure can be implemented in a recursive form. Figure 5.3 shows the diagram of the recursive version of the process.

Because of the additive noise in the detected wavefield samples, the estimate of the phase errors may not be exact. Thus, the error estimation–correction procedure can be implemented in a recursive form. Figure 5.3 shows the diagram of the recursive version of the process.

Figure 5.4 shows the result of a laboratory experiment. It shows the profile of one point target with quadrature phase error. Figure 5.5 is the estimate of the phase-error distribution. Figure 5.6a–c is the estimates of the phase errors after 1, 3, and 5 iterations, respectively. Figure 5.7 shows the gradual reduction of the quadrature phase errors after interactions.

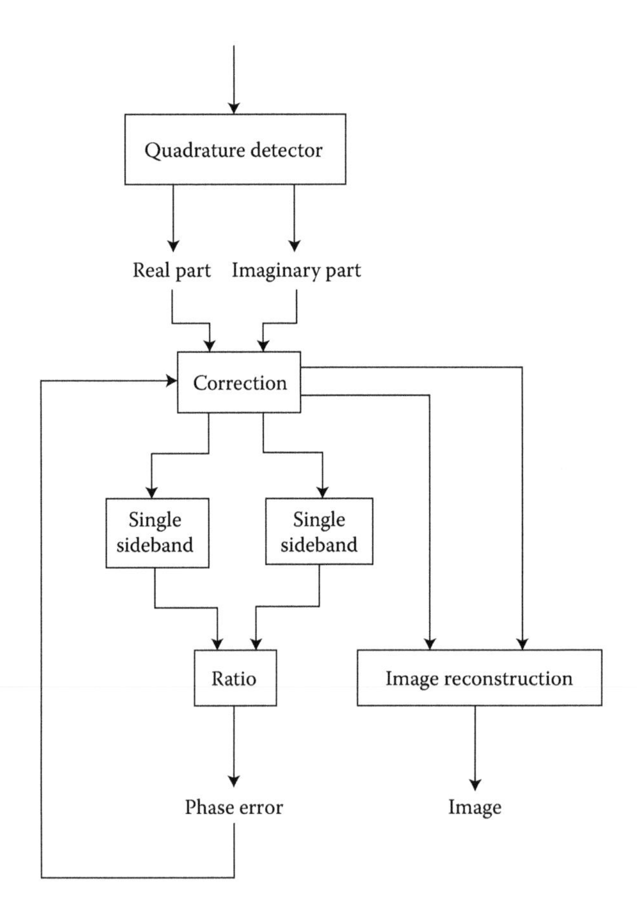

FIGURE 5.3
Recursive version of the error estimation–correction procedure.

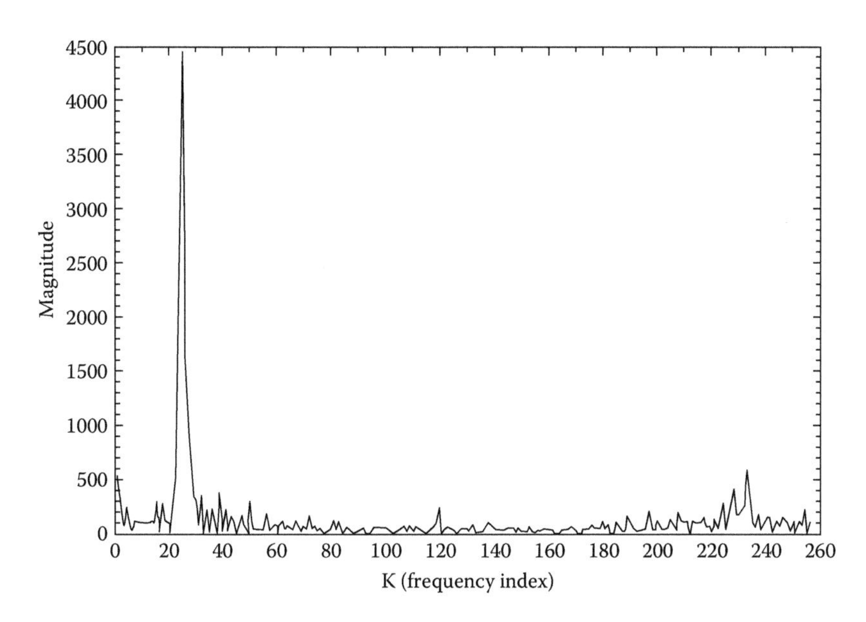

FIGURE 5.4
Profile of one point target with quadrature phase error.

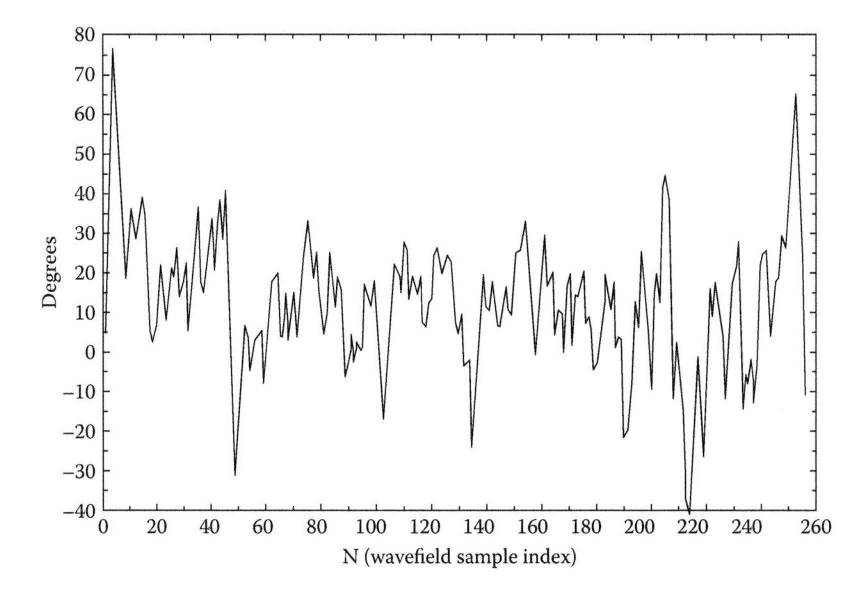

FIGURE 5.5
Initial estimate of the phase-error distribution.

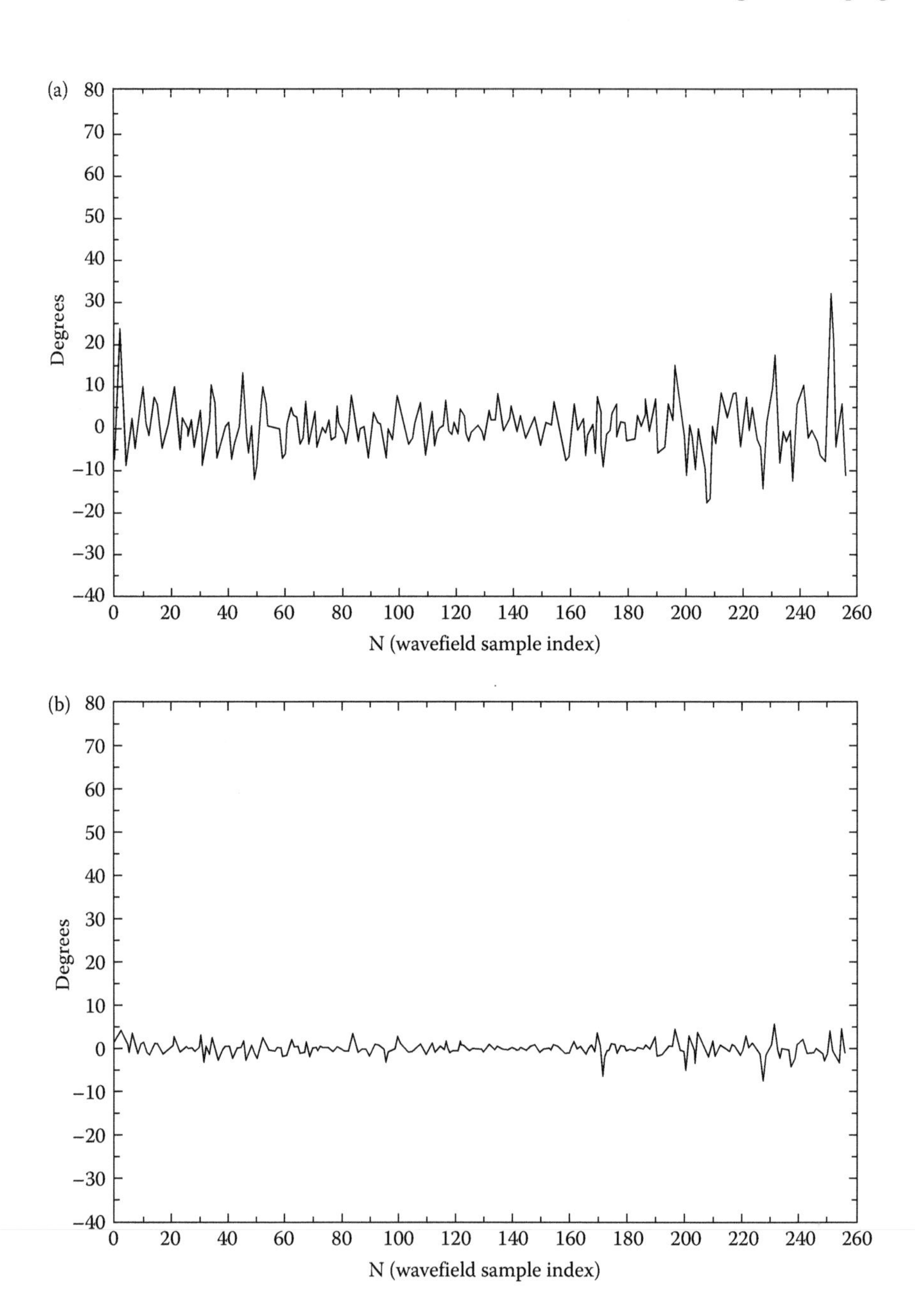

FIGURE 5.6
(a) Estimate of the phase errors after 1 iteration. (b) Estimate of the phase errors after 3 iterations. (c) Estimate of the phase errors after 5 iterations. *(Continued)*

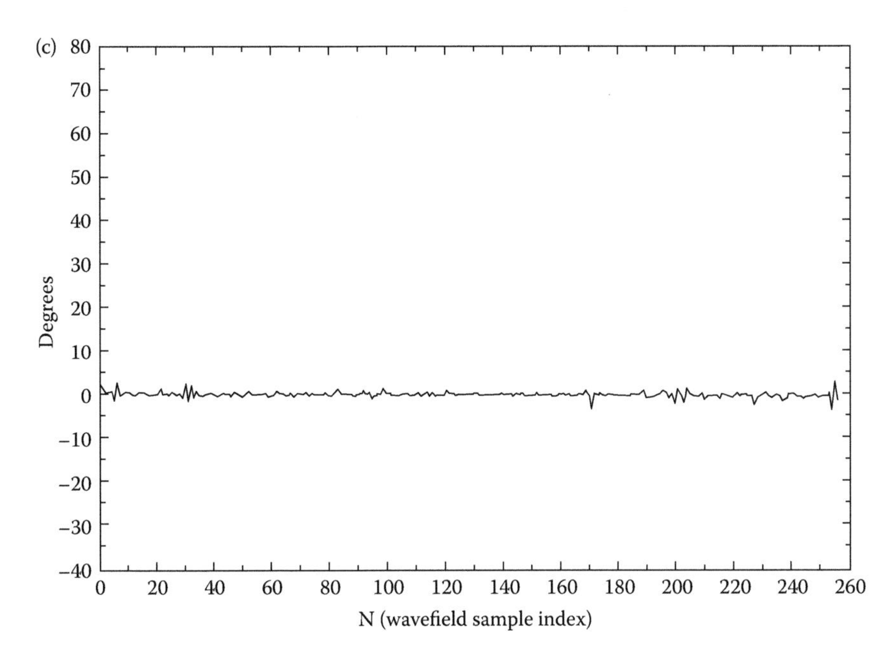

FIGURE 5.6 (Continued)
(a) Estimate of the phase errors after 1 iteration. (b) Estimate of the phase errors after 3 iterations. (c) Estimate of the phase errors after 5 iterations.

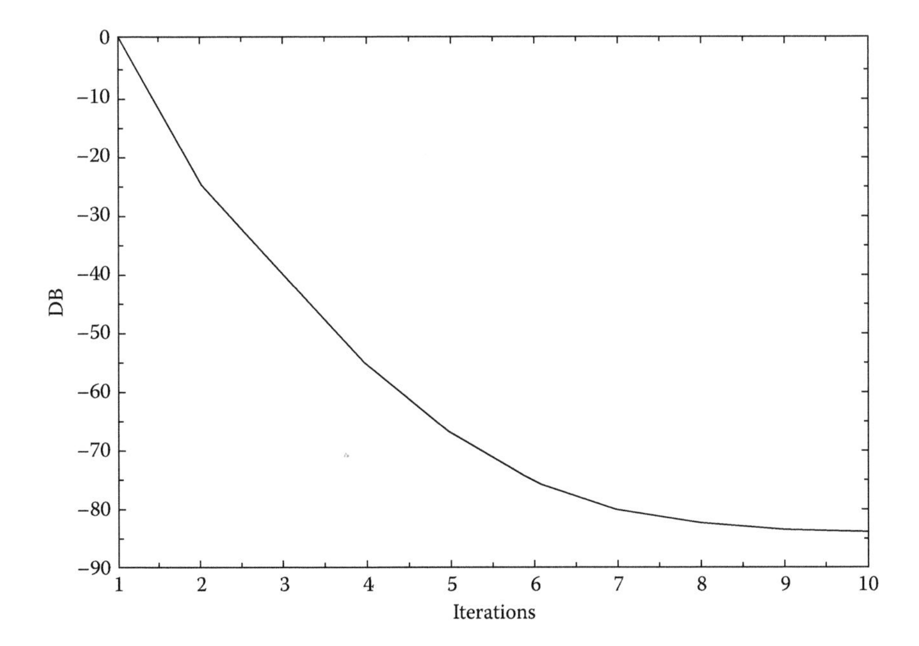

FIGURE 5.7
Gradual reduction of the quadrature phase errors after interactions.

5.2 Resolution Enhancement by Wavefield Statistics

5.2.1 Image Reconstruction by Coherent Backward Propagation

Consider a coherent source $s(x, y, z)$, the resultant wavefield $g(x, y, z)$ can be written in the form of a convolution integral over the source region S,

$$g(x,y,z) = s(x,y,z) * h(x,y,z)$$
$$= \iiint_S s(x',y',z')h(x-x',y-y',z-z')\,dx'\,dy'\,dz'$$

where $h(x, y, z)$ denotes Green's function associated with the coherent wave propagation,

$$h(x,y,z) = \left(\frac{1}{j\lambda r}\right)\exp\left(\frac{j2\pi r}{\lambda}\right)$$

and

$$r = (x^2 + y^2 + z^2)^{1/2}$$

The coherent operating wavelength is denoted as λ. The backward-propagation image formation process can also be represented as a convolution integral, where the detected wavefield $g(x, y, z)$ is convolved over the aperture R with the conjugated version of Green's function,

$$\hat{s}(x,y,z) = g(x,y,z) * h^*(x,y,z)$$
$$= \iiint_R g(x',y',z')h^*(x-x',y-y',z-z')\,dx'\,dy'\,dz'$$

where $\hat{s}(x, y, z)$ is the reconstructed image.

This linear convolution integral is commonly referred to as the backward-propagation procedure, which is equivalent to the matched filtering operation.

To visualize the image formation process, we place a point source at (x_0, y_0, z_0) as an example, for simplicity,

$$s(x,y,z) = \delta(x-x_0, y-y_0, z-z_0)$$

The wavefield produced by this point source is in the form of a spatially-shifted Green's function,

$$\begin{aligned} g(x,y,z) &= \delta(x-x_0,y-y_0,z-z_0) * h(x,y,z) \\ &= h(x-x_0,y-y_0,z-z_0) \\ &= \left(\frac{1}{j\lambda r'}\right)\exp\left(\frac{j2\pi r'}{\lambda}\right) \end{aligned}$$

where

$$r' = ((x-x_0)^2 + (y-y_0)^2 + (z-z_0)^2)^{1/2}$$

In this case, the backward-propagation image formation is now in the form of

$$\begin{aligned} \hat{s}(x,y,z) &= g(x,y,z) * h^*(x,y,z) \\ &= \iiint_R \left(\frac{1}{j\lambda r'}\right)\exp\left(\frac{j2\pi r'}{\lambda}\right)\left(\frac{-1}{j\lambda r}\right)\exp\left(\frac{-j2\pi r}{\lambda}\right) dx'\,dy'\,dz' \\ &= \iiint_R \left(\frac{1}{\lambda^2 rr'}\right)\exp\left(\frac{j2\pi(r'-r)}{\lambda}\right) dx'\,dy'\,dz' \\ &= \iiint_R A(r,r')\exp(j\theta(r,\ r'))\, dx'\,dy'\,dz' \end{aligned}$$

where

$$r = ((x-x')^2 + (y-y')^2 + (z-z')^2)^{1/2}$$

The numerical formation of the image at a particular location (x, y, z) can be regarded as a sum of a collection of vectors. The amplitude variation of these vectors is in the form of

$$A(r,r') = \frac{1}{\lambda^2 rr'}$$

and the phase is

$$\exp(j\theta(r,r')) = \exp\left(\frac{j2\pi(r'-r)}{\lambda}\right)$$

When we reach the location of the point source at $(x, y, z) = (x_0, y_0, z_0)$, the distances r and r' become identical,

$$r' = r$$

As a result, the phase part of the vectors becomes zero,

$$\exp(j\theta(r,r')) = \exp\left(\frac{j2\pi(r-r')}{\lambda}\right) = 1$$

and

$$A(r,r') = \frac{1}{\lambda^2 r^2}$$

Thus, at that location, it becomes an integral of a collection of real and positive values, which produces a large final value. For other locations, on the other hand, it remains to be a vector addition, which induces cancellation due to the variation of the phase terms, producing smaller values. This simple vector accumulation–cancellation effect is the foundation of the backward-propagation image formation process.

5.2.2 Wavefield Statistics

Here, emerges a very interesting observation. First, at a position (x, y, z) in the region of interest, the backward-propagation image is the result of the superposition of a collection of vectors.

$$\hat{s}(x,y,z) = \iiint_R g(x',y',z')h^*(x-x',y-y',z-z')\,dx'\,dy'\,dz'$$

This implies that the value of the final image is in proportion to the mean value of this vector set. Secondly, at the location of the point source, the phase terms are zero, and the vectors become real and positive scalars. This translates into smaller variance. This simple observation indicates that the statistics of the wavefield vectors consistently have large mean value and small variance at the location of the scatters. Otherwise, the statistics give small mean and large variance. That is why the variance image $v(x, y, z)$ often appears to be the inverse of the mean image $m(x, y, z)$.

Based on this observation, it has been proposed to utilize the variance to improve the resolving capability numerically in the form

$$\hat{s}'(x,y,z) = \frac{m(x,y,z)}{e(x,y,z)}$$

where $m(x, y, z)$ and $e(x, y, z)$ denote the mean value and enhancement operator of the wavefield vector set at the location (x, y, z), respectively. The enhancement operator can be written in the general form of

$$e(x,y,z) = c + dv^p(x,y,z)$$

where p is the power of the variance image $e(x, y, z)$. The constants c and d are real and positive weighting coefficients for computation stability purpose. The coefficients can be selected effectively according to the noise level.

The enhancement operator, $e(x, y, z)$, gives amplification at the scatter locations and attenuation to the background fluctuation where targets are absent. As a result, the operator provides an effective layer of the image enhancement to the backward propagated images. Figure 5.8a shows the original image of six scatters in three pairs, and Figure 5.8b is the enhanced image with first-order variance profile.

5.2.3 Wideband Modality

For wideband or multi-frequency imaging modalities, the backward-propagation image reconstruction procedure can be generalized as the superposition of a collection of coherent sub-images $\hat{s}(x, y, z; \lambda)$,

$$\hat{s}(x,y,z) = \int_{\Lambda} \hat{s}(x,y,z;\lambda)\, d\lambda$$

$$= \int_{\Lambda}\iiint_{R} \left(\frac{1}{\lambda^2 rr'}\right) \exp\left(\frac{j2\pi(r-r')}{\lambda}\right) dx'\, dy'\, dz'\, d\lambda$$

where Λ denotes the range of operating wavelengths.

Although this procedure involves multi-dimensional integration, the concept of vector accumulation–cancellation effect remains exactly the same. For each position (x, y, z) in the source region, the process provides the statistics with the mean image $m(x, y, z)$ and variance image $v(x, y, z)$. The distribution $m(x, y, z)$ is equivalent to the conventional backward-propagation image, and $v(x, y, z)$ can be used as an enhancement operator in a similar manner.

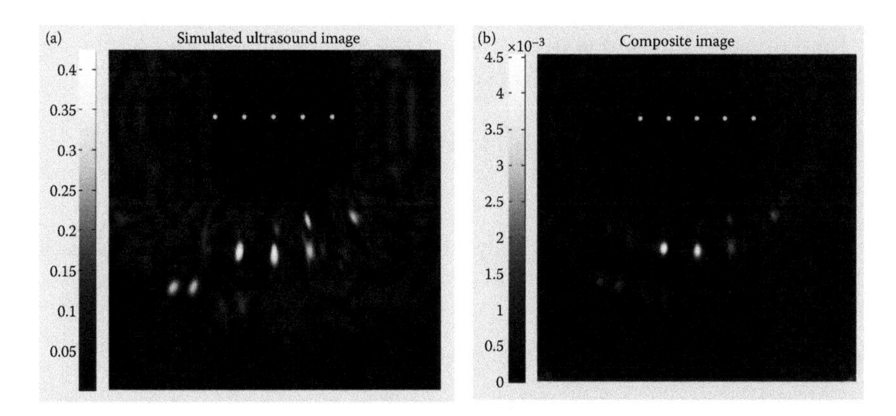

FIGURE 5.8
(a) Reconstructed image (left) and (b) enhanced image (right).

5.2.4 Cascade Form

One interesting feature is that this procedure is normally partitioned into two steps in the so-called *cascade form,* in practice. One step is the integration over the wavelength λ, and the other is an integral over the receiver aperture in the space domain.

One version is to reconstruct the coherent sub-images first, and the final image is the superposition of all coherent sub-images. This means that the first step is to form the coherent sub-images $\hat{s}(x, y, z; \lambda)$ by backward propagation in the space domain,

$$\hat{s}(x,y,z;\lambda) = \iiint_R \left(\frac{1}{\lambda^2 rr'}\right) \exp\left(\frac{j2\pi(r-r')}{\lambda}\right) dx'\,dy'\,dz'$$

Subsequently, the final image is the superposition of all the coherent sub-images,

$$\hat{s}(x,y,z) = \int_\Lambda \hat{s}(x,y,z;\lambda)\, d\lambda$$

The second version is to reverse the order of the cascade sequence. The first step is a one-dimensional integration over the wavelength. For FMCW systems, this step can be implemented in the form of integration over the frequency variable. In practice, this step is commonly referred to as the formation of the *range profiles.*

$$\hat{s}(r,r') = \int_\Lambda \left(\frac{1}{\lambda^2 rr'}\right) \exp\left(\frac{j2\pi(r-r')}{\lambda}\right) d\lambda$$

The second step is to map the range profiles to the source region to form the final image. This step is performed in the space domain, and mathematically equivalent to the backward-propagation process.

$$\hat{s}(x,y,z) = \iiint_R \hat{s}(r,r')\, dx'\,dy'\,dz'$$

For the cascade versions of the image reconstruction procedure, the calculation of the *mean* and *variance* now need to be implemented in a sequential manner.

5.2.5 Calculation of Wavefield Statistics

For simplicity, we can describe the formation of *mean*(x, y, z) and *var*(x, y, z) in the form of computing the wavefield statistics from a set of complex wavefield vectors. Assume there are M multi-static transceiver tracks and

N operating wavelengths. At each position (x, y, z) within the source region, the image $s'(x, y, z)$ is the sum of $M \times N$ complex wavefield vectors $q(x_m, y_m, z_m; \lambda_n)$ contributed from M transceivers and N operating wavelengths,

$$s'(x,y,z) = \sum_m \sum_n q(x_m, y_m, z_m; \lambda_n)$$

The mean value of these vectors is in a similar form of, off by a factor MN,

$$\begin{aligned} m(x,y,z) &= \frac{1}{MN} \sum_m \sum_n q(x_n, y_n, z_m; \lambda_n) \\ &= \frac{1}{M} \sum_m \frac{1}{N} \sum_n q(x_n, y_n, z_m; \lambda_n) \\ &= \frac{1}{N} \sum_n \frac{1}{M} \sum_m q(x_n, y_n, z_m; \lambda_n) \end{aligned}$$

This suggests the overall mean value $m(x, y, z)$ can be calculated sequentially. The sub-images are formed from the mean values of the vector subsets, and the mean value of the entire process is computed as the mean value of all the sub-images. And the overall mean value distribution $m(x, y, z)$ is the same, independent of the order of the sequential steps.

This observation is consistent to the concept of the *sum of independent random variables*, where the overall mean is the average of the means. Now, if we further extend this concept and assume the sub-images are independent observations of the source region, the variance image $v(x, y, z)$ of the complete process is the sum of the sub-images' variances.

5.3 Parameter-Based Motion Estimation

5.3.1 Point Features and Matching Correspondences

In general, the three-dimensional displacement of a rigid object contains two components. One is the translational motion, which is commonly represented by an additive vector T. The second component is the rotation, of which the operation is represented by a 3×3 orthonormal matrix $[R]$. Thus, the relationship of the coordinates of the point features of a rigid object, before and after the displacement, can be written as

$$q_i' = [R]\, q_i + T + n_i$$

where $\{q_i\}$ and $\{q_i'\}$ are the coordinates of a finite collection of point features before and after the motion, respectively, and $\{n_i\}$ denotes the measurement errors, modeled as additive noise. Note that $\{q_i\}$ and $\{q_i'\}$ are two sets of 3D position vectors. We now define two new sets of position vectors $\{p_i\}$ and $\{p_i'\}$ by removing the centroids,

$$p_i = q_i - \frac{1}{N}\sum_{i=1}^{N} q_i$$

and

$$p_i' = q_i' - \frac{1}{N}\sum_{i=1}^{N} q_i'$$

Because of the removal of the centroids, the translational vector T is totally isolated from the equation, and the linear relationship before and after the displacement is simplified down to the form of

$$[P'] = [R][P] + [n]$$

where $[P]$ is a $3 \times N$ matrix, of which the N columns are the position vectors $\{p_i\}$,

$$[P] = [\, p_1, \, p_2, \ldots, p_N]$$

and similarly, $[P']$ is the position matrix composed by the vectors $\{p_i'\}$.

$$[P'] = [p_1', p_2', \ldots, p_N']$$

Then, a matrix $[n]$ can also be formed to denote the measurement errors,

$$[n] = [n_1, n_2, \ldots, n_N]$$

Thus, after the removal of the centroids, the motion–displacement estimation is simplified to problem of estimating the least-square solution of the 3×3 rotation matrix $[R]$ given the $[P]$ and $[P']$. This means that the first step of the displacement estimation process is to estimate the orthonormal rotation matrix from the coordinates of a set of point features, before and after the displacement.

We first form a 3×3 matrix $[P'][P]^T$ and perform singular-value decomposition,

$$[P'][P]^T = [A][\Lambda][B]$$

The least-square solution is in the form of

$$[R] = [A][B]^T$$

Subsequently, the translation vector can be determined by applying the solution $[R]$ to the comparison of the centroids.

$$T = \frac{1}{N}\left[\sum_{i=1}^{N} q_i' - [R]\sum_{i=1}^{N} q_i\right]$$

Figure 5.9 shows the motion estimation procedure with point features and matching correspondences.

5.3.2 Without Matching Correspondences

Without the matching correspondences, the equation becomes

$$[P'] = [R][P][M] + [n]$$

where $[M]$ is a mutation matrix, where

$$[M][M]^T = [M]^T[M] = [I]$$

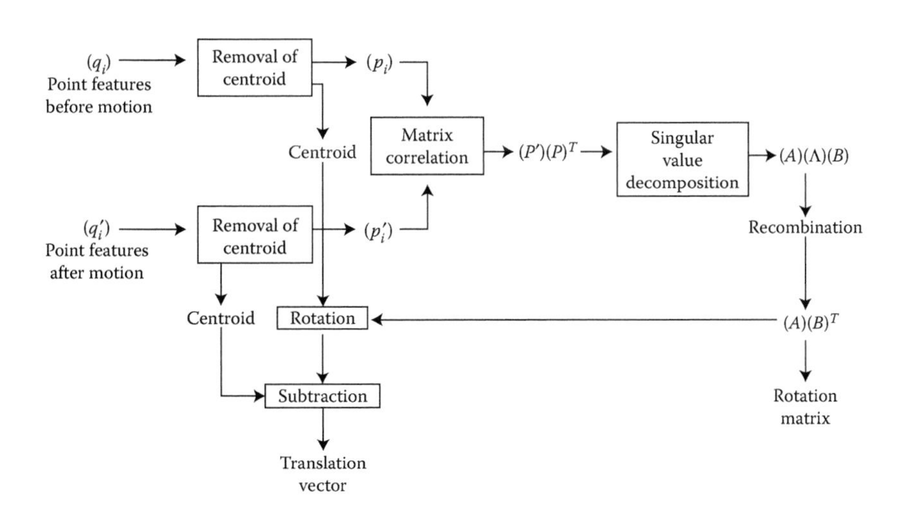

FIGURE 5.9
Diagram of the motion estimation procedure with point features and matching correspondences.

It can also be seen that, without the matching correspondences, it is not feasible to obtain the matrix $[P'][P]^T$. Thus instead, we compute $[P'][P']^T$ and $[P][P]^T$ separately instead,

$$
\begin{aligned}
[P'][P']^T &= [R][P][M][M]^T[P]^T[R]^T + [n][n]^T \\
&= [R][P][P]^T[R]^T + \rho^2[I] \\
&= [R]\{[P][P]^T + \rho^2[I]\}[R]^T
\end{aligned}
$$

Because both $[P'][P']^T$ and $[P][P]^T$ are square and symmetrical matrices, eigenvalue decomposition is performed:

$$[P][P]^T = [Q] = [U][\Lambda][U]^T$$

$$[P'][P']^T = [Q'] = [V][\Lambda'][V]^T$$

This relationship between $[Q]$ and $[Q']$ is in the simple form of

$$
\begin{aligned}
[P'][P']^T &= [Q'] = [V][\Lambda'][V]^T \\
&= [R]\{[P][P]^T + \rho^2[I]\}[R]^T \\
&= [R]\{[Q] + \rho^2[I]\}[R]^T \\
&= [R]\{[U][\Lambda][U]^T + \rho^2[I]\}[R]^T \\
&= [R][U]\{[\Lambda] + \rho^2[I]\}[U]^T[R]^T
\end{aligned}
$$

This leads to the relationship of the eigenvalues

$$[\Lambda'] = [\Lambda] + \rho^2[I]$$

and

$$[V] = [R][U]$$

The solution of the rotation matrix $[R]$ is then in the form of

$$[R] = [V][U]^T$$

And, subsequently, the translation vector $[T]$ can be calculated from the resultant rotation matrix and the centroids as before.

$$T = \frac{1}{N}\left[\sum_{i=1}^{N} q_i' - [R]\sum_{i=1}^{N} q_i\right]$$

This also indicates if the displacement contains only translational motion, the procedure degenerates into a simple form that the displacement vector can be estimated from the change of the centroids before and after the motion.

$$T = \frac{1}{N}\left[\sum_{i=1}^{N} q_i' - \sum_{i=1}^{N} q_i\right]$$

Figure 5.10 shows the diagram of the motion estimation procedure without matching correspondences.

5.3.3 Complex-Image Statistics

We denote the image before and after the displacement as $p(x)$ and $p'(x)$, respectively, where x *denotes the three-dimensional vector* $\mathbf{x} = [x, y, z]$. Then, these two images are related in the form of

$$p'(x) = p([R]^T(x - T))$$

We then utilize $p(x)$ and $p'(x)$ as the probability distributions, and we can compute the mean position of the distributions:

$$S_o = E(x) = \int xp(x)\,dx$$

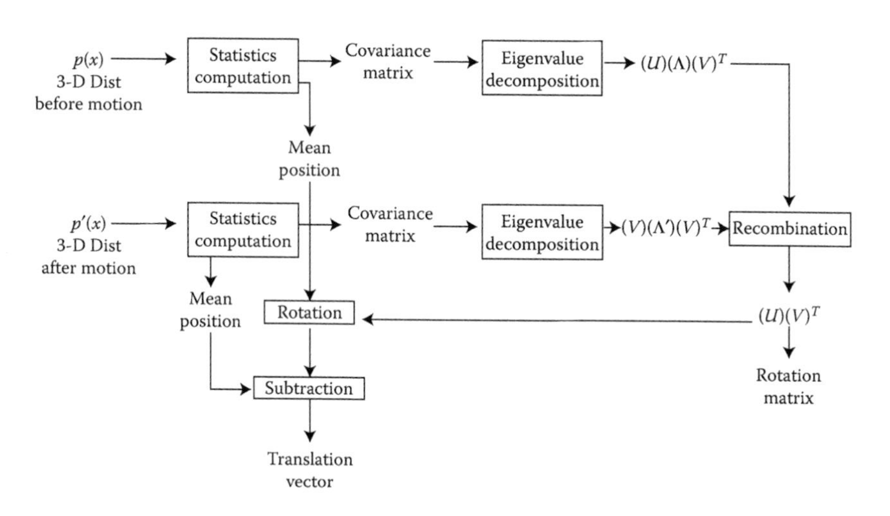

FIGURE 5.10
Motion estimation procedure without matching correspondences.

and

$$S_o' = E(x') = \int xp'(x)\, dx$$

Upon close examination, we find the same relationship between the two mean position vectors,

$$\begin{aligned} S_o' &= \int xp'(x)dx = \int xp([R]^T(x - T))\, dx \\ &= \int ([R]x + T)p(x)\, dx \\ &= [R]\int xp(x)dx + T\int p(x)\, dx \\ &= [R]S_o + T \end{aligned}$$

Then, similarly, we compute the 3×3 covariance matrices $[Q]$ and $[Q']$ as

$$\begin{aligned} [Q] &= E\{(x - S_o)(x - S_o)^T\} \\ &= \int (x - S_o)(x - S_o)^T p(x)\, dx \end{aligned}$$

and

$$\begin{aligned} [Q'] &= E\{(x - S_o')(x - S_o')^T\} \\ &= \int (x - S_o')(x - S_o')^T p'(x)\, dx \end{aligned}$$

This produces a relationship similar to the previous case,

$$\begin{aligned} [Q'] &= \int (x - S_o')(x - S_o')^T p'(x)\, dx \\ &= \int (x - S_o')(x - S_o')^T p([R]^T(x - T))\, dx \\ &= \int ([R]x + T - S_o')([R]x + T - S_o')^T p(x)\, dx \\ &= [R]\left\{\int (x - S_o)(x - S_o)^T p(x)\, dx\right\}[R]^T \\ &= [R][Q][R]^T \end{aligned}$$

As expected, we can determine the orthonormal rotation matrix in the form of

$$[R] = [V][U]^T$$

Subsequently, the translation vector can be obtained from the mean positions

$$\boldsymbol{T} = \boldsymbol{S}_o' - [R]\,\boldsymbol{S}_o$$

For translation-only movement, the rotation matrix degenerates into $[R] = [I]$, and the translational vector can be simplified to the differential of the centroids,

$$\boldsymbol{T} = \boldsymbol{S}_o' - \boldsymbol{S}_o$$

This technique allows us to effectively perform 3D motion estimation from sequences of acoustical images directly, without the need for point features and matching correspondences. This approach can be further enhanced by conducting the motion estimation in the spatial-frequency domain.

In the frequency domain, the translation vector results in linear phase shift of the spectrum. And, because of the rotation-invariant property of the Fourier transform, the rotation component of the 3D motion lies in the magnitude distributions. Thus, the magnitude and phase information of the complex-image profiles can be partitioned and used for the estimation of the rotation matrix and translation vector separately. Figure 5.11 shows the diagram of the motion estimation procedure with complex-image distributions in the spatial-frequency domain.

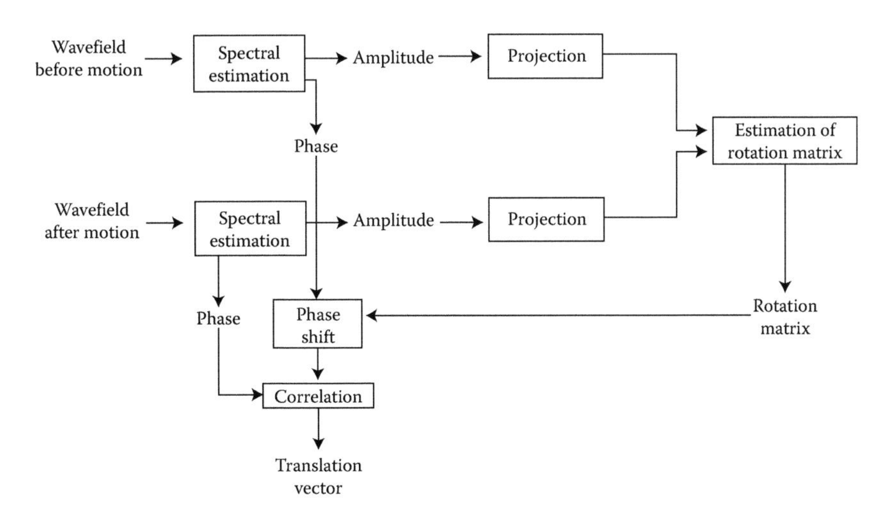

FIGURE 5.11
Motion estimation procedure with complex-image distributions.

5.4 Image-Based Motion Estimation and Imaging

5.4.1 Image-Domain Methods

Typical motion detection is to compare the adjacent image frames to visualize the changes. This is known as the first-order differential, which is the most basic method,

$$\hat{p}_n(x,y,z) = s_n(x,y,z) - s_{n-1}(x,y,z)$$

where $s_n(x, y, z)$ denote the image sequence in time. The image sequence $\hat{p}_n(x, y, z)$ shows the changes. To give image changes with smoother differentials, higher-order procedures are often applied. The second-order and third-order procedures are, respectively,

$$\hat{p}_n(x,y,z) = s_n(x,y,z) - 2\, s_{n-1}(x,y,z) + s_{n-2}(x,y,z)$$

and

$$\hat{p}_n(x,y,z) = s_n(x,y,z) - 3s_{n-1}(x,y,z) + 3s_{n-2}(x,y,z) - s_{n-3}(x,y,z)$$

Time-differential technique is a widely used approach and is efficient in terms of computation complexity. This technique is effective for motion estimation without target rotation.

5.4.2 Data-Domain Methods

Because both the differential-detection operator and image formation algorithm are linear operations, the order of these two operations can be reversed and still achieve the same result. Thus, we can apply the differential operation to examine the wavefield data, in the form of

$$\hat{g}_n(x,y,z) = g_n(x,y,z) - g_{n-1}(x,y,z)$$

Similarly, higher-order differential operations can be applied and the second-order and third-order cases are in the form of

$$\hat{g}_n(x,y,z) = g_n(x,y,z) - 2g_{n-1}(x,y,z) + g_{n-2}(x,y,z)$$

and

$$\hat{g}_n(x,y,z) = g_n(x,y,z) - 3g_{n-1}(x,y,z) + 3g_{n-2}(x,y,z) - g_{n-3}(x,y,z)$$

Separation of the stationary and varying components in the data domain has one most significant advantage, which is the compression of wavefield data and simplification of computation. Event-triggered sensing and imaging approach is largely based on this simple concept.

5.4.3 Hybrid Version

The concept outlined in this section is based on the linearity and shift invariance of the motion estimation/detection step and the image formation procedure. For that, the result is the same, completely independent of the order of the procedures. The motion-detection step can be executed before or after the image formation process.

As discussed in the previous chapters, the image formation procedure for multi-frequency imaging modality contains two components. The first version of the image formation algorithm is to produce a collection of coherent sub-images, followed by a superposition process. The second version produces a collection of sub-images from the range profiles first, and the combination gives the final image. For both versions, it is feasible as well as interesting to place the motion-detection step in between. In this section, one of the two versions is selected to illustrate the concept of this hybrid procedure.

Consider the case of placing the displacement detection step between the range-profile estimation and superposition process. For each transmitter–receiver pair, a range profile $p(r)$ is constructed. For continuing monitoring over a period of time, the data-acquisition process provides a sequence $\{p_n(r)\}$. If we conduct the change detection at this stage, the first-order version is in the form of a pair of sequences $\{\hat{p}_n(r)\}$ and $\{\hat{q}_n(r)\}$.

$$\hat{q}_n(r) = \frac{1}{2}(p_n(r) - p_{n-1}(r))$$

$$\hat{p}_n(r) = \frac{1}{2}(p_n(r) + p_{n-1}(r))$$

The sequence $\{\hat{q}_n(r)\}$ represents the change of the range profile. The sequence $\{\hat{p}_n(r)\}$ is corresponding to the stationary component. The summation of these two sequences recovers the original

$$\hat{p}_n(r) + \hat{q}_n(r) = p_n(r)$$

Higher-order versions of this procedure can also be used for this purpose. If the target distribution is stationary, the sequence $\hat{q}_n(r)$ becomes zero.

This suggests that the separation enables us to produce to separate images. The image produced from the sequence $\{\hat{p}_n(r)\}$ is the image of the stationary profile, and the image from $\{\hat{q}_n(r)\}$ shows the time-varying component.

5.4.4 Frequency-Domain Analysis

The most interesting exercise is to examine this simple relationship in the frequency domain. If we Fourier transform the range profile to bring it to the frequency domain, it is in the form of

$$P_n(f) = F\{p_n(r)\}$$

Consider the simple case of one single target. If the target produces a displacement due to motion, the range profile is a shifted version

$$p_{n+1}(r) = p_n(r - \Delta r_n)$$

In the frequency domain, the relationship becomes

$$P_{n+1}(f) = P_n(f)\exp(-j2\pi f \Delta r_n)$$

The most challenging and interesting application of the motion estimation techniques is the case of microperiodic motions. Because of the small magnitude of the displacement, the perturbation is within the resolution cell size. As a result, the range profiles remain unchanged and the final images from this time sequence do not show sufficient changes to detect the motion or displacement. Thus, the traditional space-domain motion estimation methods are not effective. To formulate the analysis, the displacement can be modeled in a simple form of

$$\Delta r_n = \Delta r \sin(n\Omega_0)$$

where Ω_0 is the angular increment of the oscillatory movement and Δr is the magnitude of the oscillation. It should be noted that the value of Δr is small. Now, if we approximate the phase term as

$$\exp(-j2\pi f \Delta r_n) \approx 1 - j2\pi f \Delta r_n$$

then, the range profile becomes

$$\begin{aligned} P_n(f) &= P(f)\exp(-j2\pi f \Delta r_n) \\ &= P(f)\exp(-j2\pi f \Delta r \sin(n\Omega_0)) \\ &\approx P(f)[1 - j2\pi f \Delta r \sin(n\Omega_0)] \end{aligned}$$

Further consolidation brings it to

$$\begin{aligned} P(f,n) &= P_n(f) \\ &= P(f)[1 - \pi f \Delta r \exp(jn\Omega_0) + \pi f \Delta r \exp(-jn\Omega_0)] \end{aligned}$$

If an N-point FFT is applied for each frequency f, it contains three peaks

$$P(f,k) = P(f)\delta(k) - \pi f \Delta r P(f)\delta(k - N\Omega_0/2\pi) + \pi f \Delta r P(f)\delta(k + N\Omega_0/2\pi)$$

Because of the oscillatory movement, the FFT spectrum shows two extra peaks at

$$k = \pm N\Omega_0/2\pi$$

From the locations of the peaks, we can determine the frequency of the oscillation. It can also be seen that the frequency shift is uniquely related to the frequency. Thus, if we partition the distribution $P(f, k)$ according to the frequency index, we can reconstruct time-varying component of the distribution accordingly. By comparing the magnitudes of the peaks, we can also determine the magnitude of the oscillatory movement.

The following experiment involves two targets with microoscillatory motion at different frequency. Figure 5.12 is the traditional image profiles. Figure 5.13 is the frequency spectrum of the range-profile sequence, which shows two point targets at different range distances with different oscillatory frequencies.

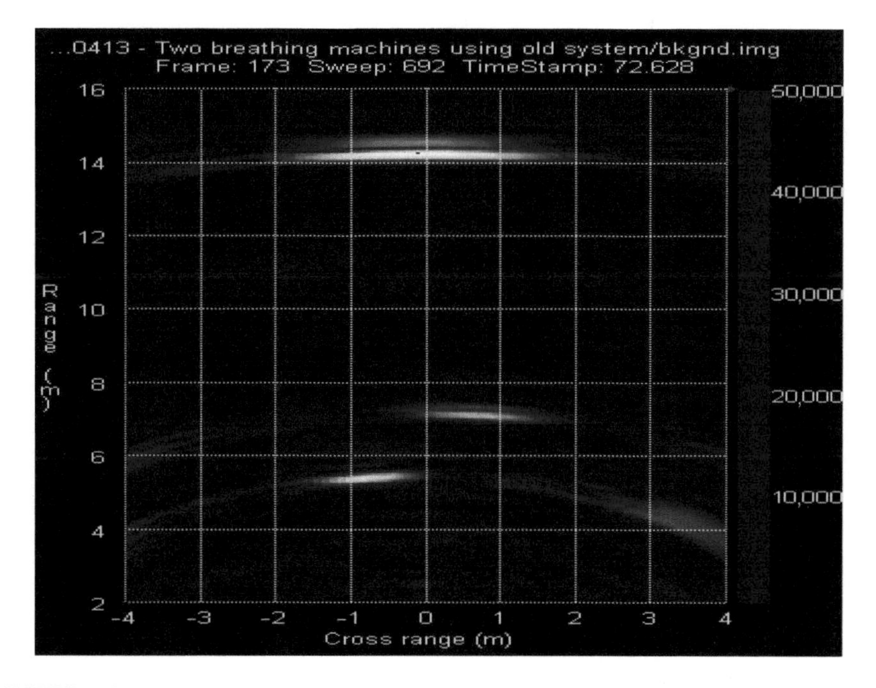

FIGURE 5.12
Traditional image profiles of two moving targets.

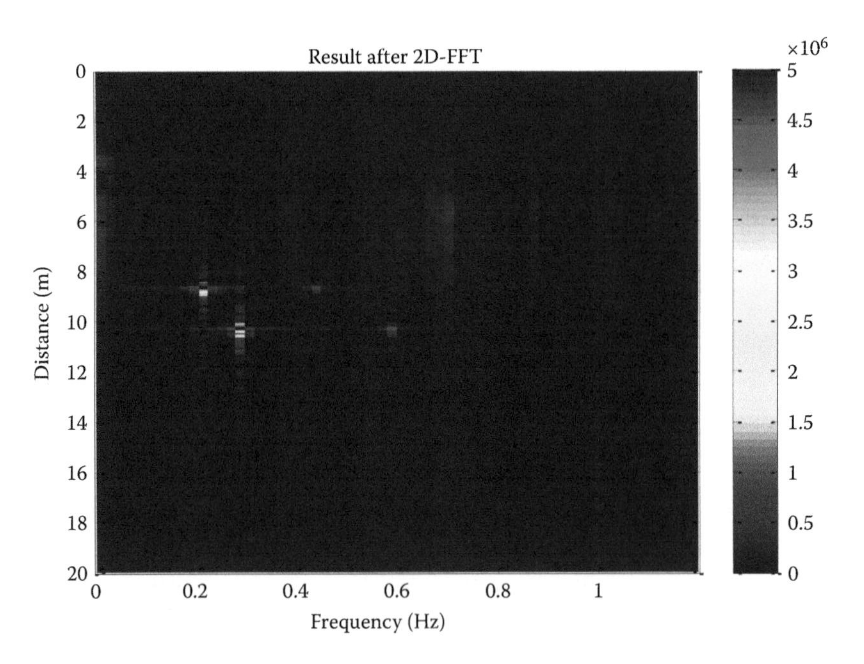

FIGURE 5.13
Frequency spectrum of the range-profile sequence.

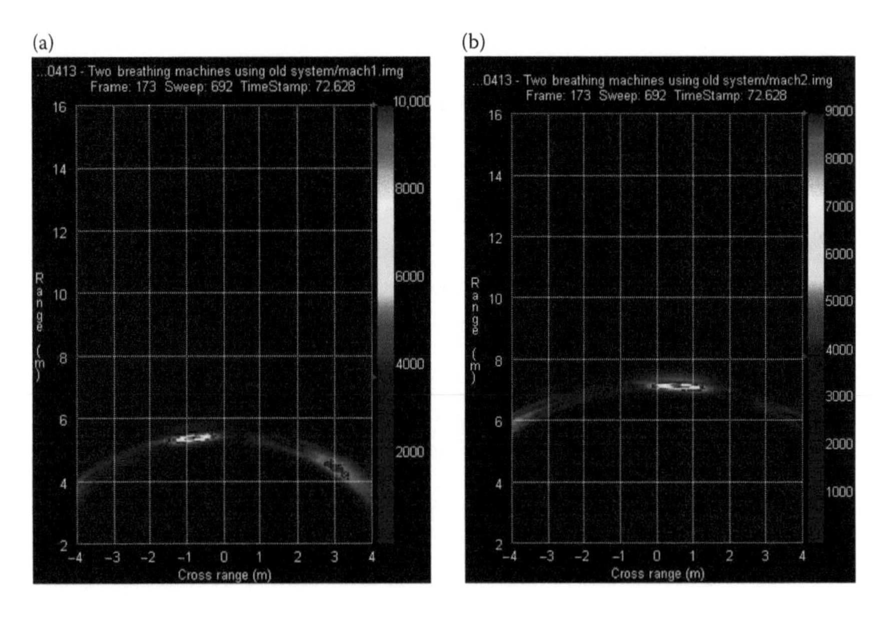

FIGURE 5.14
Reconstructed image of moving target (a) and target (b).

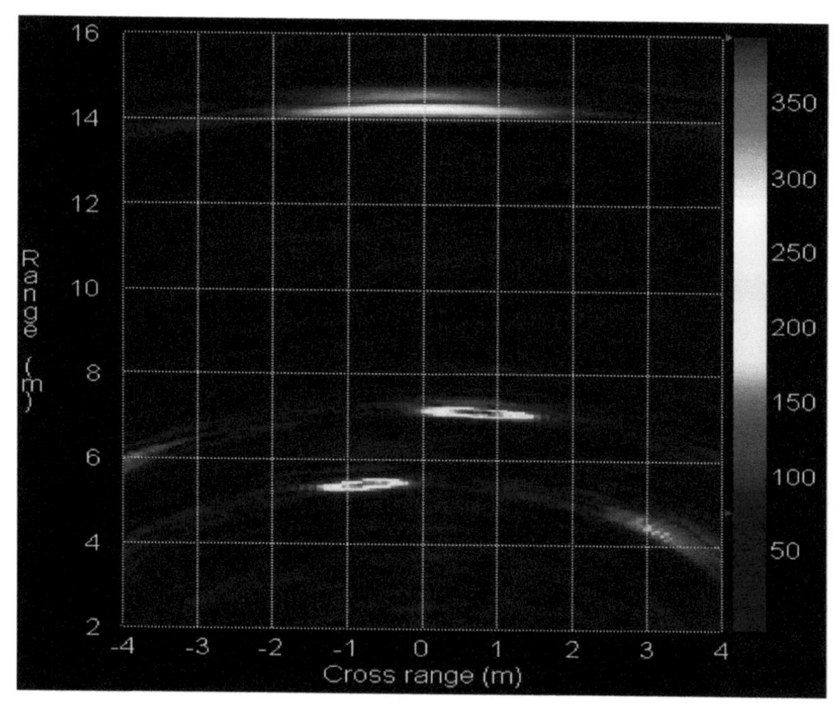

FIGURE 5.15
Superimposed image, with time-varying components in color and stationary components shown in gray scale.

We can then partition the range profiles according to the frequency index. This allows us to reconstruct two separate images of the two targets corresponding to their oscillatory frequencies. Figure 5.14a and b shows the reconstructed images of the two moving targets, respectively.

The time-varying components of the image can be placed upon the overall image for the improvement of the quality of visualization. Figure 5.15 shows the superimposed image, with time-varying components in color and stationary background shown in gray scale.

Index